CISM COURSES AND LECTURES

Series Editors:

The Rectors of CISM
Sandor Kaliszky - Budapest
Mahir Sayir - Zurich
Wilhelm Schneider - Wien

The Secretary General of CISM
Giovanni Bianchi - Milan

Executive Editor
Carlo Tasso - Udine

The series presents lecture notes, monographs, edited works and proceedings in the field of Mechanics, Engineering, Computer Science and Applied Mathematics.
Purpose of the series is to make known in the international scientific and technical community results obtained in some of the activities organized by CISM, the International Centre for Mechanical Sciences.

INTERNATIONAL CENTRE FOR MECHANICAL SCIENCES

COURSES AND LECTURES - No. 371

NON LINEAR ANALYSIS AND BOUNDARY VALUE PROBLEMS FOR ORDINARY DIFFERENTIAL EQUATIONS

EDITED BY

F. ZANOLIN
UNIVERSITY OF UDINE

Springer-Verlag Wien GmbH

Le spese di stampa di questo volume sono in parte coperte da
contributi del Consiglio Nazionale delle Ricerche.

This volume contains 1 illustrations

In order to make this volume available as economically and as
rapidly as possible the authors' typescripts have been
reproduced in their original forms. This method unfortunately
has its typographical limitations but it is hoped that they in no
way distract the reader.

ISBN 978-3-211-82811-3   ISBN 978-3-7091-2680-6 (eBook)
DOI 10.1007/978-3-7091-2680-6

PREFACE

The theory of ordinary differential equations has received a strong impulse by the introduction of some methods from Nonlinear Analysis (like, e.g., fixed point theorems, degree theory, topological invariants, variational methods) and the study of boundary value problems for ODEs is, at present, a rich area full of significant achievements, both from the theoretical and the applied sides. This volume concerns the study of some boundary value problems for nonlinear ordinary differential equations, using topological or variational methods, with a special emphasis also on the so-called qualitative (or geometric) approach in the theory of ordinary differential equations, making use of a broad collection of techniques coming from Analysis and Topology which are addressed to the study of ODEs.

In this book, the study of more classical boundary value problems for ordinary differential equations (like the periodic or the Sturm - Liouville ones) which represent the main interest of a wide number of researchers in the world, is accompanied by the presentation of some recent results dealing with other interesting and more atypical BVPs. In particular, there are new contributions to the study of bounded solutions and to the presentation of some very recent approaches yielding to a rigorous proof of the presence of chaotic motions. The aspect of numerical treatment of dynamical systems is discussed as well and some new achievements about the validity of discretization methods are analyzed.

The volume is addressed to people interested in Nonlinear Analysis, Ordinary Differential Equations, or Boundary Value Problems. In particular, various items chosen from the following areas are included: Topological degree and related topics; Fixed point theorems and applications; Bifurcation theory; Application of variational methods; Qualitative theory of ODEs; Topological invariants for the study of dynamical systems; Chaotic motions; Bounded solutions; Numerical methods for the analysis of dynamical systems.

Thanks to CISM hospitality, it was possible to organize in Udine a School on Nonlinear Analysis and Boundary Value Problems for Ordinary Differential Equations. Financial support was given by various different sources. In particular, we mention UNESCO-ROSTE which provided some scholarships to partially support participants from developing countries and Eastern Europe. We are also grateful to other international, national and local institutions, like the European Community (through grants CI1*-CT93-0323 and ERB CHRX-CT94-0555), the C.N.R. (Italy), the MURST (project "Equazioni Differenziali Ordinarie e Applicazioni", by professor Roberto Conti), the University of Udine with the Department of Mathematics and Computer Science and the CRUP Bank.

The School consisted of a series of lectures delivered by some well-known specialists, devoted to focus the basic facts and the recent advances of the theory as well as to present interesting research problems. We had also a few one-hour invited lectures which were addressed to more specific topics. Then, the rest of the time was devoted to a number of communications and seminars concerning the presentation of recent research achievements.

The success of the School was due to the kindness of all these distinguished colleagues who accepted to share their mathematical expertise with others and delivered some highly interesting lectures, as well as to the active presence of all the participants. We are also indebted to professors A. Capietto, A. Fonda, M. Gaudenzi, M. Marini and P. Omari who helped in various important steps for the organization of the Conference. We finally thank professor Carlo Tasso and the CISM staff for the publication of these proceedings.

This volume collects six contributions, prepared by expert and active mathematicians, corresponding to the content of their CISM lectures. Special thanks are due to Jean Mawhin who found the right words to commemorate Gilles Fournier, a friend and colleague who made brilliant contributions in this area of research. This book is thus dedicated to the memory of Gilles.

F. Zanolin

IN MEMORIAM GILLES FOURNIER

Before starting a school devoted to topological methods in nonlinear boundary value problems, we cannot avoid remembering the recent untimely death of one of the most active and original contributors to this area of mathematics.

Gilles Fournier died unexpectedly on August 14 1995, leaving his beloved wife Reine and their three children André, Marie-Hélène and Nadine in deep sorrow.

A student at the University of Montréal, where he got his PhD in 1973 under the direction of Andrzej Granas, Gilles Fournier has made his whole career at the University of Sherbrooke, being an indefatigable animator of its Department of Mathematics and Computer Science, organizing several international conferences and directing many PhD dissertations.

After some important work in algebraic topology, and specially in fixed point theory, Gilles Fournier became interested in nonlinear differential equations, using his talent in topology to develop new tools for proving the existence and multiplicity of solutions.

An enthousiastic traveller, Gilles Fournier had in particular many contacts with Italian mathematicians, spending several months as visiting professor at the University of Calabria.

Gilles Fournier's extraordinary intuition and constant enthusiasm has led him to joint work with many colleagues, some of them present here. Through his important mathematical legacy and the memory of his sociable and friendly personality, Gilles Fournier will stay in our mind and in our heart, always remaining the gifted and cheerful companion we all remember.

Jean Mawhin

CONTENTS

UPPER AND LOWER SOLUTIONS IN THE THEORY OF
ODE BOUNDARY VALUE PROBLEMS:
CLASSICAL AND RECENT RESULTS

C. De Coster and P. Habets
Catholic University of Louvain, Louvain-la-Neuve, Belgium

1   Introduction

The method of upper and lower solutions for ordinary differential equation was introduced in 1931 by G. Scorza Dragoni for a Dirichlet problem. Since then a large number of contributions enriched the theory. Among others, one has to point out the pioneer work of M. Nagumo who associated his name with derivative dependent right hand side.

Basically, the method of upper and lower solutions deals with existence results for boundary value problems. These upper and lower solutions can be thought of as numerical approximations of solutions that satisfy the equations up to an error term with constant sign. The method describes problems so that existence of a solution is inferred from two such approximations with error terms of opposite sign.

In this set of notes, we consider two such problems: the periodic one and Dirichlet boundary value problem. We limit ourself to second order ordinary differential equations and only consider problems with derivative independent right-hand side. This last restriction wipes out the important problem of computing a-priori bounds for the derivative of the solutions.

At first, the method of upper and lower solutions might look as being mainly of theoretical interest since it takes for granted the existence of such functions. This believe cuts off the large number of applications based on the method. It is true that in practical situations it is often difficult to recognize that the assumptions at hand imply the existence of such upper and lower solutions and we must admit that to exhibit such functions is somewhat of an art. For this reason, a large part of this text

---

*Chargée de recherches du fonds national belge de la recherche scientifique.

is devoted to applications and we believe that it is the practice of such problems which really gives the ability and skill necessary to use the full power of the method.

The first section of these notes is of general character. We present the notion of $W^{2,1}$-upper and lower solutions which is adapted to prove existence of solutions in the $W^{2,1}$ sense. This notion allows also angles in the graph of these functions. This aims to deal with applications such as singular perturbation problems. The Dirichlet problem is studied for systems with singularities at the boundary points. Applications to mechanical problems with singular forces and to Landesman-Lazer conditions are worked out. Relation with degree theory is used to deal with multiplicity results and is applied to the pendulum equation. A multiplicity result is also described using the connection between the variational approach and the method of upper and lower solutions. At last this introductory section ends with an introduction on monotone iterative schemes related to upper and lower solutions.

Systems with singularities are investigated in the second section. The problem of interest concerns existence of positive solutions for a Dirichlet problem which is singular both at the end points of the time interval and at $u = 0$. Such situations appear in applied mathematics problems such as the Emden-Fowler equation. In the second part of this section, existence of pairs of positive solutions are considered.

An Ambrosetti-Prodi problem with two parameters is considered in Section 4.

The next section is devoted to upper and lower solutions in the reversed order. This type of result is worked out for the periodic problem. A first result concerns upper and lower solutions without ordering. In the second part, we consider systems with asymmetric nonlinearities and assume the upper and lower solutions are ordered. The last result concerns the use of monotone methods for systems with upper and lower solutions in the reversed order.

Historical and bibliographical notes are given in the last section.

Throughout the text, some proofs are left as exercises, extensions are presented as problems. The interested reader might find there the opportunity of further readings and a chance to develop his ability to work upper and lower solutions.

The study made here is far from being complete. We can of course consider other applications. We can extend the class of boundary value problem we consider or deal with partial differential equations. Upper and lower solutions can be used to obtain bounded solutions or homoclinics, higher order equations and systems can be considered. Also, upper and lower solutions are efficient in other instances such as stability problems. These subjects would be a natural complement of our work.

Definitions and notations

A function $f(t, u)$ defined on $E \subset [a, b] \times \mathbb{R}$ is said to be a *Carathéodory function* or to satisfy *Carathéodory conditions on E* if
  (i) for almost every $t \in [a, b]$, $f(t, \cdot)$ is continuous on its domain;
  (ii) for any $u \in \mathbb{R}$, the function $f(\cdot, u)$ is measurable on its domain.
A function $f(t, u)$ defined on $E \subset [a, b] \times \mathbb{R}$ is said to be an *L^p-Carathéodory function,*

$p \geq 1$, if it is a Carathéodory function and $\forall r > 0$, $\exists h_r \in L^p(a, b)$, $\forall u \in [-r, r]$ and for a.e. $t \in [a, b]$ with $(t, u) \in E$, $|f(t, u)| \leq h_r(t)$.

"for a.e." means "for almost every" and "a.e." means "almost everywhere".

$D_- u(t)$, $D^- u(t)$, $D_+ u(t)$, and $D^+ u(t)$ are the Dini derivatives of the continuous function u (see, for example, E.J. McShane [97]).

$D_l u(t)$ and $D_r u(t)$ are the left and right derivatives of u.

$x \prec y$ means $\exists \epsilon > 0$, $\forall t \in [a, b]$, $y(t) - x(t) \geq \epsilon \sin\left(\pi \frac{t-a}{b-a}\right)$.

$u^+ := \max(u, 0)$

$u^- := \max(-u, 0)$

$\mathcal{A} := \{h \in L^1_{loc}(]0, \pi[) \mid s(\pi - s)h(s) \in L^1(0, \pi)\}$

$C^k(I, J)$ is the set of k times continuously differentiable functions $f : I \to J$

$C^k(I) := C^k(I, \mathbb{R})$

$C^k_0(I, J) := \{u \in C^k(I, J) \mid I = [a, b], u(a) = 0, u(b) = 0\}$

$C^k_0(I) := C^k_0(I, \mathbb{R})$

$C_0(I) := C^0_0(I)$

$C_{2\pi} := \{u \in C(\mathbb{R}) \mid \text{for all } t \in \mathbb{R}, u(t) = u(t + 2\pi)\}$

$H^k(I) := W^{k,p}(I)$

$H^1_0(I) := H^1(I) \cap C_0(I)$

$L^p(I) := \{u : I \to \mathbb{R} \text{ measurable} \mid \int_I |u(t)|^p dt < +\infty\}$, where $p \in [1, +\infty[$

$L^p_{loc}(I) := \{u : I \to \mathbb{R} \text{ measurable} \mid \forall J \subset I \text{ compact}, u \in L^p(J)\}$

$L^\infty(I) := \{u : I \to \mathbb{R} \text{ measurable} \mid \exists R > 0, \text{ for a.e. } t \in I, |u(t)| \leq R\}$

$W^{k,p}(I) := \{u \in C^{k-1}(I) \mid u^{(k)} \in L^p(I)\}$

$W^{2,\mathcal{A}}(0, \pi) := \{u \in W^{1,1}(0, \pi) \mid \ddot{u} \in \mathcal{A}\}$

$\|u\|_{C^k} := \max_{i=1,\ldots,k}\{\sup_{t \in I} |u^{(i)}(t)|\}$

$\|u\|_\infty := \| \cdot \|_{C^0}$

$\|u\|_{L^p} := (\int_I |u(t)|^p dt)^{1/p}$

2 The method of upper and lower solutions

The first part of this chapter is devoted to existence results for the periodic and Dirichlet boundary value problem. We consider both C^2 and $W^{2,1}$-solutions. In each case, we tried to introduce notions of upper and lower solutions which are general enough to deal with most applications and to characterize the essential of the method. We did not try to encompass all possible definitions. For example, we consider lower and upper solutions with angles in order to deal with applications such as singular perturbation problems. To work out such definitions displays that the method is based on a maximum principle. In general, this amounts to an extremum criteria based on the second derivative but, in case of angles, it is achieved by first order conditions.

As a selection of possible applications we consider periodic solutions of problems with singular forces and Landesman-Lazer conditions for Dirichlet problem. These examples illustrate the fact that upper and lower solutions are related to nonlinearity "on the left of the first eigenvalue".

In Section 2.3, we concentrate on the relation with degree theory. We want to associate, to the fixed point operator T associated with the BVP and to the set Ω of functions which are between the upper and lower solutions, a degree. This imposes there is no fixed point of T on $\partial\Omega$ and motivates the introduction of a stronger notion of upper and lower solutions. Also, it imposes to choose carefully the space on which we work. A basic result is that this degree is always equal to 1. Existence of degree leads to multiplicity results of two different types. The first one requires the existence of two pairs of strict lower and upper solutions and can be thought of as a "local" result. It will be illustrate by a pendulum type result. The other one requires, beside the existence of a pair of strict lower and upper solutions, an a priori bound, i.e. an assumption at infinity.

Relation between upper and lower solutions and variational method is considered in Section 2.4. We present there a multiplicity result for Dirichlet problem due to P. Omari and F. Zanolin [104].

The last section presents rough ideas on the relation between upper and lower solutions and the monotone method. Here we can compute approximations of the solutions from converging sequences of upper and lower solutions.

2.1   Second Order Periodic BVP

Existence of C^2-solutions

Consider the periodic problem

$$\ddot{u} + f(t, u) = 0,$$
$$u(0) = u(2\pi), \quad \dot{u}(0) = \dot{u}(2\pi), \tag{2.1}$$

where f is a continuous function. The simplest approach to the method of lower and upper solutions for second order ordinary differential equations uses the following definition.

Definition 2.1 *A function $\alpha \in C^2(]0, 2\pi[) \cap C^1([0, 2\pi])$ is a* lower solution *of the periodic problem* (2.1) *if :*
(a) for all $t \in]0, 2\pi[$, $\ddot{\alpha}(t) + f(t, \alpha(t)) \geq 0$;
(b) $\alpha(0) = \alpha(2\pi)$, $\dot{\alpha}(0) \geq \dot{\alpha}(2\pi)$.
   A function $\beta \in C^2(]0, 2\pi[) \cap C^1([0, 2\pi])$ is an upper solution *of the problem* (2.1) *if*

(a) for all $t \in]0, 2\pi[$, $\ddot{\beta}(t) + f(t, \beta(t)) \leq 0$;
(b) $\beta(0) = \beta(2\pi)$, $\dot{\beta}(0) \leq \dot{\beta}(2\pi)$.

The main result of the method is an intermediate value theorem. It proves that if we can find a lower solution which is smaller than an upper one, there is a solution wedged between these two.

Theorem 2.1 *Let α and β be lower and upper solutions of (2.1) such that $\alpha \leq \beta$. Define*

$$E = \{(t, u) \in [0, 2\pi] \times \mathbb{R} \mid \alpha(t) \leq u \leq \beta(t)\}$$

and assume f is continuous on E. Then the problem (2.1) has at least one solution $u \in C^2([0, 2\pi])$ such that, for all $t \in [0, 2\pi]$,

$$\alpha(t) \leq u(t) \leq \beta(t).$$

Proof : Consider the modified problem

$$\ddot{u} - u + f(t, \gamma(t, u)) + \gamma(t, u) = 0,$$
$$u(0) = u(2\pi), \quad \dot{u}(0) = \dot{u}(2\pi), \tag{2.2}$$

where $\gamma : [0, 2\pi] \times \mathbb{R} \to \mathbb{R}$ is defined by

$$\begin{aligned} \gamma(t, u) &= \alpha(t), && \text{if } u < \alpha(t), \\ &= u, && \text{if } \alpha(t) \leq u \leq \beta(t), \\ &= \beta(t), && \text{if } u > \beta(t). \end{aligned} \tag{2.3}$$

First we prove that all solutions u of (2.2) satisfy

$$\alpha(t) \leq u(t) \leq \beta(t).$$

Let us assume on the contrary that, for some $t_0 \in [0, 2\pi]$,

$$\min_t(u(t) - \alpha(t)) = u(t_0) - \alpha(t_0) < 0.$$

If $t_0 \in {]0, 2\pi[}$, we obtain the contradiction

$$0 \leq \ddot{u}(t_0) - \ddot{\alpha}(t_0) = -f(t_0, \alpha(t_0)) + u(t_0) - \alpha(t_0) - \ddot{\alpha}(t_0) < 0.$$

In case the minimum is achieved at the boundary points,

$$\min_t(u(t) - \alpha(t)) = u(0) - \alpha(0) = u(2\pi) - \alpha(2\pi) < 0,$$

we obtain

$$\dot{u}(0) - \dot{\alpha}(0) \geq 0 \geq \dot{u}(2\pi) - \dot{\alpha}(2\pi),$$

and, by the definition of a lower solution, $\dot{u}(0) - \dot{\alpha}(0) \leq \dot{u}(2\pi) - \dot{\alpha}(2\pi)$. Hence, we have $\dot{u}(0) - \dot{\alpha}(0) = 0$ and for $t > 0$ small

$$\dot{u}(t) - \dot{\alpha}(t) = \int_0^t [-f(s, \alpha(s)) + u(s) - \alpha(s) - \ddot{\alpha}(s)] ds < 0$$

which is a contradiction. This proves $\alpha(t) \leq u(t)$.

In a similar way, we prove $u(t) \leq \beta(t)$.

Next, we write (2.2) as an integral equation

$$u(t) = \int_0^{2\pi} G(t, s)[f(s, \gamma(s, u(s))) + \gamma(s, u(s))]ds,$$

where $G(t, s)$ is the Green function corresponding to the problem

$$\ddot{u} - u + f(t) = 0,$$
$$u(0) = u(2\pi), \ \dot{u}(0) = \dot{u}(2\pi). \tag{2.4}$$

The operator

$$T : \mathcal{C}([0, 2\pi]) \to \mathcal{C}([0, 2\pi])$$

defined by

$$(Tu)(t) = \int_0^{2\pi} G(t, s)[f(s, \gamma(s, u(s))) + \gamma(s, u(s))]ds$$

is completely continuous and bounded. By Schauder's theorem, T has a fixed point which is a solution of (2.2) and, from the first part of the proof, it is also a solution of (2.1). ∎

Remark 2.1 Observe that this result gives us two kind of information. It is an existence result but we also have a localization of the solution which can be very useful as shown in the next examples.

Example 2.1 Consider

$$\epsilon^2 \ddot{u} - u^3 + \sin^3 t = 0,$$
$$u(0) = u(2\pi), \ \dot{u}(0) = \dot{u}(2\pi),$$

where $\epsilon > 0$ is a small parameter. It is easy to see that

$$\alpha(t) = \sin t - 2\epsilon \text{ and } \beta(t) = \sin t + 2\epsilon$$

are lower and upper solutions. Hence, there is a solution

$$u(t) = \sin t + O(\epsilon).$$

Notice that beside existence, the upper and lower solutions give an asymptotic estimate on the solution.

Example 2.2 Consider the problem

$$\ddot{u} + \sin u = h(t),$$
$$u(0) = u(2\pi), \ \dot{u}(0) = \dot{u}(2\pi),$$

where $h \in C([0, 2\pi])$. Then, if $\|h\|_\infty \leq 1$ there is a solution u such that

$$\frac{\pi}{2} \leq u(t) \leq \frac{3\pi}{2}. \tag{2.5}$$

Observe that, if h is constant, the set of solutions which satisfy (2.5) contains the unstable equilibrium. This result is general; the upper and lower solution method exhibits unstable solutions.

Remark 2.2 Let us notice first that Theorem 2.1 is not valid if $\alpha \geq \beta$. Consider for example the problem

$$\ddot{u} + u = \sin t,$$
$$u(0) = u(2\pi), \quad \dot{u}(0) = \dot{u}(2\pi).$$

This problem has no solution although $\alpha(t) = 1$ and $\beta(t) = -1$ are lower and upper solutions.

Also, the result is no more true if we reverse the inequalities (b) in Definition 2.1. Consider for example the problem

$$\ddot{u} = 1,$$
$$u(0) = u(2\pi), \quad \dot{u}(0) = \dot{u}(2\pi).$$

Here there is no solution although $\alpha(t) = (t-\pi)^2 - 1$ and $\beta(t) = \pi^2$ satisfy $\ddot{\alpha}(t) = 2 > 1$, $\ddot{\beta}(t) \leq 1$ and $\beta(t) > \alpha(t)$. The theorem does not apply since $\dot{\alpha}(0) < \dot{\alpha}(2\pi)$.

Problem 2.1 (Alternative proof) Assume $\alpha < \beta$ and prove the previous result from the homotopy

$$\ddot{u} + \lambda f(t, u) - (1 - \lambda) \left[k^2 \left(u - \frac{\alpha(t) + \beta(t)}{2} \right) + \frac{\ddot{\alpha}(t) + \ddot{\beta}(t)}{2} \right] = 0,$$
$$u(0) = u(2\pi), \quad \dot{u}(0) = \dot{u}(2\pi),$$

where k is large enough. Use degree theory and the set

$$\Omega = \{ u \in C([0, 2\pi]) \mid \forall t \in [0, 2\pi], \ \alpha(t) < u(t) < \beta(t) \}$$

(see C. Fabry - P. Habets [47]).

Existence of $W^{2,1}$-solutions

Considering the proof of Theorem 2.1, the argument is different if the minimum of $u - \alpha$ is achieved at $t_0 \in {]}0, 2\pi{[}$ or at $t_0 = 0$. In this last case, we used an integral argument. This can be generalized and is the key to existence of $W^{2,1}$-solutions.

In this section, we consider the problem

$$\ddot{u} + f(t, u) = 0,$$
$$u(0) = u(2\pi), \quad \dot{u}(0) = \dot{u}(2\pi), \tag{2.6}$$

where f is a "Carathéodory" function . To simplify the notations, we extend $f(t, u)$ by periodicity $f(t, u) = f(t + 2\pi, u)$ and consider the set

$$C_{2\pi} = \{ u \in C(\mathbb{R}) \mid \text{ for all } t \in \mathbb{R}, \ u(t) = u(t + 2\pi) \}.$$

Definition 2.2 *A function $\alpha \in C_{2\pi}$ is a $W^{2,1}$-lower solution of (2.6) if, for any $t_0 \in \mathbb{R}$, either $D_-\alpha(t_0) < D^+\alpha(t_0)$, or there exists an open interval I_0 such that $t_0 \in I_0$, $\alpha \in W^{2,1}(I_0)$ and, for almost every $t \in I_0$,*

$$\ddot{\alpha}(t) + f(t, \alpha(t)) \geq 0.$$

In the same way, a function $\beta \in C_{2\pi}$ is a $W^{2,1}$-upper solution of (2.6) if, for any $t_0 \in \mathbb{R}$, either $D^-\beta(t_0) > D_+\beta(t_0)$, or there exists an open interval I_0 such that $t_0 \in I_0$, $\beta \in W^{2,1}(I_0)$ and, for almost every $t \in I_0$,

$$\ddot{\beta}(t) + f(t, \beta(t)) \leq 0.$$

When there is no possible confusion, we will, as previously, speak of lower and upper solutions of (2.6).

Remark 2.3 Observe that, if we extend by periodicity α and β to all of $\mathbb{R}$, the boundary conditions in Definition 2.1 give the same kind of corners as those allowed in Definition 2.2.

Remark 2.4 If f is continuous and $\alpha \in C^1([0, 2\pi]) \cap C^2(]0, 2\pi[)$, this function is a $W^{2,1}$-lower solution if and only if it satisfies Definition 2.1. A similar statement holds for upper solutions.

Remark 2.5 If the lower solutions α_1 and α_2 intersect a finite number of times, we can prove (see [33]) the maximum $\max(\alpha_1, \alpha_2)$ is a lower solution. However, we do not know if the result holds in full generality.

Remark 2.6 Definition 2.2 is related to several definitions that were introduced in the literature. A selection of such definitions can be found in P. Habets and L. Sanchez [62], J. J. Nieto [100], P. Habets and M. Laloy [60], C. Fabry and P. Habets [48], M. Nagumo [99], J. Deuel and P. Hess [43], D. G. de Figueiredo and S. Solimini [42] and J. Mawhin [90].

Theorem 2.2 *Let α and β be $W^{2,1}$-lower and upper solutions of (2.6) such that $\alpha \leq \beta$. Define $E = \{(t, u) \in [0, 2\pi] \times \mathbb{R} \mid \alpha(t) \leq u \leq \beta(t)\}$ and assume $f : E \to \mathbb{R}$ is a L^1-Carathéodory function.*

Then the problem (2.6) has at least one solution $u \in W^{2,1}(0, 2\pi)$ such that, for all $t \in [0, 2\pi]$,

$$\alpha(t) \leq u(t) \leq \beta(t).$$

Proof : As in the proof of Theorem 2.1 we consider the modified problem (2.2) where γ is defined by (2.3). It is easy to adapt the proof of this theorem to show that problem (2.2) has a solution and we only have to prove that this solution satisfies

$$\alpha(t) \leq u(t) \leq \beta(t).$$

Let us assume on the contrary that, for some $t_0 \in \mathbb{R}$,

$$\min_t(u(t) - \alpha(t)) = u(t_0) - \alpha(t_0) < 0.$$

Then we have

$$\dot{u}(t_0) - D_-\alpha(t_0) \leq \dot{u}(t_0) - D^+\alpha(t_0).$$

By the definition of a $W^{2,1}$-lower solution we have equality, there exists an open interval I_0 with $t_0 \in I_0$, $\alpha \in W^{2,1}(I_0)$ and, for almost every $t \in I_0$,

$$\ddot{\alpha}(t) + f(t, \alpha(t)) \geq 0.$$

Further $\dot{u}(t_0) - \dot{\alpha}(t_0) = 0$ and, for $t \geq t_0$,

$$\dot{u}(t) - \dot{\alpha}(t) = \int_{t_0}^t (\ddot{u}(s) - \ddot{\alpha}(s))ds$$

$$\leq \int_{t_0}^t [-f(s, \alpha(s)) + u(s) - \alpha(s) - \ddot{\alpha}(s)]ds < 0.$$

This proves $u(t_0) - \alpha(t_0)$ is not a minimum of $u - \alpha$ which is a contradiction.   ∎

The following elementary example illustrates the interest of Theorem 2.2. It exemplifies a class of problems where "angular" lower and upper solutions are useful.

Example 2.3 Consider the problem

$$\epsilon^2 \ddot{u} = \varphi(u) - \varphi(|t - \pi|),$$
$$u(0) = u(2\pi), \quad \dot{u}(0) = \dot{u}(2\pi)$$

where $\varphi \in C^1(\mathbb{R})$ is such that $\varphi'(t) \geq a^2$. Notice that the functions

$$\alpha_1(t) = -(t - \pi) - \frac{\epsilon}{a} e^{-\frac{at}{\epsilon}} \quad \text{and} \quad \alpha_2(t) = (t - \pi) - \frac{\epsilon}{a} e^{\frac{a(t - 2\pi)}{\epsilon}}$$

are such that

$$\epsilon^2 \ddot{\alpha}_1(t) - \varphi(\alpha_1(t)) + \varphi(|t - \pi|) \geq -\epsilon a e^{-\frac{at}{\epsilon}} + a^2(|t - \pi| + t - \pi + \frac{\epsilon}{a} e^{-\frac{at}{\epsilon}})$$
$$\geq 0$$

and

$$\epsilon^2 \ddot{\alpha}_2(t) - \varphi(\alpha_2(t)) + \varphi(|t - \pi|)$$

$$\geq -\epsilon a e^{\frac{a(t - 2\pi)}{\epsilon}} + a^2(|t - \pi| - (t - \pi) + \frac{\epsilon}{a} e^{\frac{a(t - 2\pi)}{\epsilon}}) \geq 0.$$

Moreover the function $\alpha \in C([0, 2\pi])$ defined by $\alpha(t) = \max(\alpha_1(t), \alpha_2(t))$ is a $W^{2,1}$-lower solution. Next we can check that the function

$$\beta(t) = |t - \pi| + \frac{\epsilon}{a} e^{-\frac{a|t - \pi|}{\epsilon}}$$

is a $W^{2,1}$-upper solution. From Theorem 2.2, we deduce there exists a solution u such that, for all $t \in [0, 2\pi]$, $\alpha(t) \le u(t) \le \beta(t)$ and, from the choice of the lower and upper solutions, we obtain the asymptotic estimate

$$u(t) = |t - \pi| + O(\epsilon e^{-\frac{a|t-\pi|}{\epsilon}}).$$

Singular Forces

Consider the model example

$$\ddot{u} + \frac{1}{u} = h(t),$$
$$u(0) = u(2\pi), \quad \dot{u}(0) = \dot{u}(2\pi),$$

(2.7)

where $h \in L^1(0, 2\pi)$. Integrating the equation, it is easy to see that a necessary condition to have a positive solution of (2.7) is that

$$\overline{h} = \frac{1}{2\pi} \int_0^{2\pi} h(t) \, dt > 0.$$

(2.8)

Using lower and upper solutions, we can prove the following result.

Proposition 2.3 Let $h \in L^1(0, 2\pi)$ satisfy (2.8) and assume there exists $M \in \mathbb{R}$ such that, for a.e. $t \in [0, 2\pi]$, $h(t) \le M$. Then the problem (2.7) has at least a positive solution.

Proof : Every constant $\alpha \in]0, 1/M]$ is a lower solution. To construct the upper solution, let $\tilde{h}(t) = h(t) - \overline{h}$ and define w to be any solution of

$$\ddot{u} = \tilde{h}(t),$$
$$u(0) = u(2\pi), \quad \dot{u}(0) = \dot{u}(2\pi).$$

It is then easy to prove that, for C large enough, $\beta(t) := C + w(t)$ is an upper solution and $\beta \ge \alpha$. Hence, the result follows from Theorem 2.2. ∎

Notice that the method does not apply in case of repulsive force such as in the problem

$$\ddot{u} - \frac{1}{u} = h(t),$$
$$u(0) = u(2\pi), \quad \dot{u}(0) = \dot{u}(2\pi).$$

Attractive forces have been studied by P. Habets and L. Sanchez [62], see also A. C. Lazer - S. Solimini [82], where more results can be found.

2.2   The Dirichlet Boundary Value Problem

General Result

This section concerns the boundary value problem

$$\ddot{u} + f(t, u) = 0,$$
$$u(0) = 0, \ u(\pi) = 0. \qquad (2.9)$$

Definition 2.3 *A function* $\alpha \in C([0, \pi])$ *is a* $W^{2,1}$-*lower solution of (2.9) (or simply a lower solution) if*
(a) *for any* $t_0 \in]0, \pi[$, *either* $D_-\alpha(t_0) < D^+\alpha(t_0)$, *or there exists an open interval* $I_0 \subset]0, \pi[$ *such that* $t_0 \in I_0$, $\alpha \in W^{2,1}(I_0)$ *and, for a.e.* $t \in I_0$,

$$\ddot{\alpha}(t) + f(t, \alpha(t)) \geq 0;$$

(b) $\alpha(0) \leq 0, \ \alpha(\pi) \leq 0.$
    In the same way, a function $\beta \in C([0, \pi])$ *is a* $W^{2,1}$-*upper solution of (2.9) if,*
(a) *for any* $t_0 \in]0, \pi[$, *either* $D^-\beta(t_0) > D_+\beta(t_0)$, *or there exists an open interval* $I_0 \subset]0, \pi[$ *such that* $t_0 \in I_0$, $\beta \in W^{2,1}(I_0)$ *and, for a.e.* $t \in I_0$,

$$\ddot{\beta}(t) + f(t, \beta(t)) \leq 0;$$

(b) $\beta(0) \geq 0, \ \beta(\pi) \geq 0.$

Remark 2.7 If f is continuous and $\alpha \in C([0, \pi]) \cap C^2(]0, \pi[)$, this function is a $W^{2,1}$-lower solution if and only if
(a) for any $t \in]0, \pi[$, $\ddot{\alpha}(t) + f(t, \alpha(t)) \geq 0$;
(b) $\alpha(0) \leq 0, \ \alpha(\pi) \leq 0.$
A similar statement holds for $W^{2,1}$-upper solutions.

As in the periodic case, if f is a L^1-Carathéodory function, we can state a result which gives the existence of a solution $u \in W^{2,1}(0, \pi)$.

Theorem 2.4 *Let* α *and* β *be* $W^{2,1}$-*lower and upper solutions of (2.9) such that* $\alpha \leq \beta$. *Define* $E = \{(t, u) \in [0, \pi] \times \mathbb{R} \mid \alpha(t) \leq u \leq \beta(t)\}$ *and assume* $f : E \to \mathbb{R}$ *is a* L^1-*Carathéodory function.*
    Then the problem (2.9) has at least one solution $u \in W^{2,1}(0, \pi)$ *such that, for all* $t \in [0, \pi]$,

$$\alpha(t) \leq u(t) \leq \beta(t).$$

Exercise 2.2 Prove the preceding result using the argument of Theorem 2.2.

A natural extension concerns other boundary value problems.

Problem 2.3 Extend Theorem 2.4 to the separated boundary value problem

$$\ddot{u} + f(t, u) = 0,$$
$$a_1 u(a) - a_2 \dot{u}(a) = A,$$
$$b_1 u(b) + b_2 \dot{u}(b) = B,$$

in case $A, B \in \mathbb{R}$, $a_1, b_1 \in \mathbb{R}$, $a_2, b_2 \in \mathbb{R}^+$, $a_1^2 + a_2^2 > 0$ and $b_1^2 + b_2^2 > 0$.
Hint : See [33], [15].

An other type of extension concerns the following problem which generalizes the classical p-Laplacian.

Problem 2.4 Extend Theorem 2.4 to

$$(\phi(u'))' + f(t, u) = 0,$$
$$u(0) = 0, \ u(\pi) = 0,$$

where $\phi : \mathbb{R} \to \mathbb{R}$ is an increasing homeomorphism such that $\phi(0) = 0$.
Hint : See C. De Coster [34].

Singular problem

Applications lead to problems where the function f is singular. Consider for example the problem

$$\ddot{u} - \frac{1}{t(\pi-t)} = 0,$$
$$u(0) = 0, \ u(\pi) = 0,$$
$$(2.10)$$

which has a solution $u(t) = \frac{t}{\pi} \log \frac{t}{\pi} + \frac{\pi-t}{\pi} \log \frac{\pi-t}{\pi}$. This solution is not in $W^{2,1}(0, \pi)$ and it is clear that Theorem 2.2 does not apply. The key to singular problems such as (2.10), requires an analysis of the corresponding linear problem.

Let

$$\mathcal{A} = \{ h \in L^1_{loc}(]0, \pi[) \mid s(\pi - s)h(s) \in L^1(0, \pi) \},$$

define the set $W^{2,\mathcal{A}}(0, \pi)$ of functions in $W^{1,1}(0, \pi)$ with weak second derivative in $\mathcal{A}$ and notice that

$$W^{2,\mathcal{A}}(0, \pi) = \{ u \in W^{1,1}(0, \pi) \mid \ddot{u} \in \mathcal{A} \} \subset C([0, \pi]) \cap C^1(]0, \pi[).$$

Notice that the solution of (2.10) is in $W^{2,\mathcal{A}}(0, \pi)$. This is generalized in the following lemma.

Lemma 2.5 *If $h \in \mathcal{A}$, the problem*

$$\ddot{u} + h(t) = 0,$$
$$u(0) = 0, \ u(\pi) = 0$$
$$(2.11)$$

has a solution $u \in W^{2,\mathcal{A}}(0, \pi)$ such that

$$u(t) = \int_0^\pi G(t, s) \, h(s) \, ds,$$

where $G(t, s)$ is the Green function corresponding to (2.11).

Remark 2.8 Let $h \in L^1_{loc}(0, \pi)$. The solution of (2.11) is of the form

$$u(t) = A + Bt + \int_{\pi/2}^{t} (t - s)h(s) \, ds.$$

It is easy to see then that if we want solutions to be continuous on $[0, \pi]$, we must impose

$$\int_{0}^{\pi} t(\pi - t)h(t) \, dt < +\infty.$$

We can now state the main result for a singular problem of the form (2.9).

Theorem 2.6 *Let $D \subset]0, \pi[\times \mathbb{R}$ and let $f : D \rightarrow \mathbb{R}$ satisfy a Carathéodory condition. Assume that α and β are $W^{2,1}$-lower and upper solutions such that, for any $t \in [0, \pi]$, $\alpha(t) \leq \beta(t)$.*
Let $E = \{(t, u) \mid t \in]0, \pi[, \ \alpha(t) \leq u \leq \beta(t)\} \subset D$ and assume that there exists a function $h \in \mathcal{A}$ such that, for all $(t, u) \in E$,

$$|f(t, u)| \leq h(t). \tag{2.12}$$

Then the problem (2.9) has at least one solution $u \in W^{2,\mathcal{A}}(0, \pi)$ such that, for all $t \in [0, \pi]$,

$$\alpha(t) \leq u(t) \leq \beta(t).$$

Exercise 2.5 Prove the previous result.
Hint : See P. Habets - F. Zanolin [63], [64].

Remark 2.9 Observe that $L^1(0, \pi) \subset \mathcal{A}$, so that assumption (2.12) generalizes the classical L^1-Carathéodory conditions on f.

Remark 2.10 The condition $h \in \mathcal{A}$ is used to prove

$$\left\| \int_{0}^{\pi} G(t, s) \, h(s) \, ds \right\|_{\infty} < +\infty \text{ and } \int_{0}^{\pi} \left| \frac{\partial G}{\partial t}(t, s) \right| h(s) \, ds \in L^1(0, \pi), \tag{2.13}$$

where $G(t, s)$ is the Green function associated to (2.11). We can see that in the periodic case, the set of functions $h :]0, \pi[\rightarrow \mathbb{R}^+$ measurable that satisfy (2.13) is in fact $L^1(0, \pi)$. Hence, Theorem 2.2 has the same generality as the counterpart of Theorem 2.6.

Example 2.4 Consider the boundary value problem

$$\ddot{u} + |u|^{1/2} - \tfrac{1}{t} = 0,$$
$$u(0) = 0, \ u(\pi) = 0.$$

It is easy to see that $\beta(t) = 0$ is an upper solution and $\alpha(t) = t \ln \frac{t}{\pi} - t$ is a lower solution. Hence we have a solution u such that, for all $t \in [0, \pi]$,

$$t \ln \frac{t}{\pi} - t \leq u(t) \leq 0.$$

Observe that, in this example, the function $f(t, u)$ is not L^1-Carathéodory.

Non-resonance Conditions

Upper and lower solutions are associated with systems which can be thought of as being on the left of the first eigenvalue. The following result describes a boundary value problem

$$\ddot{u} + u + f(t, u) + h(t) = 0,$$
$$u(0) = 0, \ u(\pi) = 0,$$

(2.14)

which is asymptotically of this type.

Theorem 2.7 *Assume that* $f : [0, \pi] \times \mathbb{R} \to \mathbb{R}$ *is a* L^1*-Carathéodory function,* $h \in L^1(0, \pi)$ *and there exists* $\gamma \in L^1(0, \pi)$ *such that*

$$\limsup_{|u| \to +\infty} \frac{f(t, u)}{u} \leq \gamma(t) \lneq 0, \ uniformly \ in \ t.$$

Then the problem (2.14) has at least one solution.

Proof : Let λ_1 be the first eigenvalue of the problem

$$\ddot{u} + (1 + \gamma(t)) u + \lambda u = 0,$$
$$u(0) = 0, \ u(\pi) = 0.$$

Observe that $\lambda_1 > 0$ (see [86] together with [50]). By hypothesis, there exists $a \in L^1(0, \pi)$ such that, for all $u \geq 0$ and a.e. $t \in [0, \pi]$,

$$f(t, u) \leq (\gamma(t) + \frac{\lambda_1}{2}) u + a(t).$$

We shall choose an upper solution of the form

$$\beta(t) = w(t) + s\psi(t) \geq 0.$$

To this end, we compute

$$\ddot{\beta} + \beta + f(t, \beta) + h(t) \leq \ddot{\beta} + (1 + \gamma(t) + \frac{\lambda_1}{2}) \beta + a(t) + h(t)$$

$$\leq \ddot{w} + (1 + \gamma(t) + \frac{\lambda_1}{2}) w + a(t) + h(t) + s[\ddot{\psi} + (1 + \gamma(t) + \frac{\lambda_1}{2}) \psi]$$

As $\lambda_1/2$ is at the left of the first eigenvalue, we can choose w such that

$$\ddot{w} + (1 + \gamma(t) + \tfrac{\lambda_1}{2}) w + a(t) + h(t) = 0,$$
$$w(0) = 0, \ w(\pi) = 0$$

and $\psi(t)$ solution of

$$\ddot{\psi} + (1 + \gamma(t) + \tfrac{\lambda_1}{2}) \psi = 0,$$
$$\psi(0) = 0.$$

Such a ψ can be chosen positive as follows from a Sturm comparison argument. Hence, if s is large enough, $\beta(t)$ is a positive upper solution.

In the same way we construct a lower solution $\alpha \leq 0$.   ∎

Landesman-Lazer Conditions

The classical Landesman-Lazer conditions apply to a boundary value problem which is asymptotically resonant. In case this condition refers to a nonlinearity on the left of the first eigenvalue, upper and lower solutions are build into the problem just as in the previous case.

Theorem 2.8 *Let* $\varphi(t) = \sqrt{\frac{2}{\pi}} \sin t$. *Assume that* f *is a* L^1-*Carathéodory function,* $h \in L^1(0, \pi)$ *and*
(i) *there exist* $s_+ \in \mathbb{R}$, $f_+ \in L^1(0, \pi)$ *such that, if* $u \geq s_+ \varphi(t)$ *we have*

$$f(t, u) \leq f_+(t)$$

   and

$$\overline{f}_+ := \int_0^\pi f_+(t)\, \varphi(t)\, dt \leq -\overline{h} := -\int_0^\pi h(t)\, \varphi(t)\, dt;$$

(ii) *there exist* $s_- \in \mathbb{R}$, $f_- \in L^1(0, \pi)$ *such that, if* $u \leq s_- \varphi(t)$ *we have*

$$f(t, u) \geq f_-(t)$$

   and

$$\overline{f}_- := \int_0^\pi f_-(t)\, \varphi(t)\, dt \geq -\overline{h}.$$

Then the problem (2.14) *has at least one solution.*

Proof : Let us construct a lower solution $\alpha(t) \leq s_- \varphi(t)$. Define w to be the solution of

$$\ddot{w} + w + f_-(t) + h(t) - (\overline{f}_- + \overline{h})\varphi(t) = 0,$$
$$w(0) = 0, \ w(\pi) = 0,$$

which satisfies $\int_0^\pi w(t)\, \varphi(t)\, dt = 0$. Notice that we can find $a < 0$ small enough so that $a\, \varphi(t) + w(t) \leq s_- \varphi(t)$. Take $\alpha(t) = a\, \varphi(t) + w(t)$ and observe it is a lower solution.
   In the same way, we construct an upper solution $\beta(t) \geq s_+ \varphi(t)$. As we have chosen $\alpha(t) \leq 0 \leq \beta(t)$, we have a solution of (2.14) by application of Theorem 2.4.   ■

   In practical situations, the application of Theorem 2.8 is not straightforward. Consider, for example, the problem (2.14) in case

$$f(t, u) = -\arctan u.$$

More workable are the classical Landesman-Lazer conditions which we can deduce from Theorem 2.8.

Theorem 2.9 *Let* $\varphi(t) = \sqrt{\frac{2}{\pi}} \sin t$. *Assume that* f *is a* L^1-*Carathéodory function,* $h \in L^1(0, \pi)$ *and there exist* $\hat{f}_+, \hat{f}_- \in L^1(0, \pi)$ *such that*
(i) $\limsup\limits_{u \to +\infty} f(t, u) \leq \hat{f}_+(t)$ *uniformly in* t *and*

$$\overline{\hat{f}}_+ := \int_0^\pi \hat{f}_+(t)\, \varphi(t)\, dt < -\overline{h} := -\int_0^\pi h(t)\, \varphi(t)\, dt;$$

(ii) $\liminf\limits_{u\to-\infty} f(t,u) \geq \hat{f}_-(t)$ *uniformly in t and*

$$\overline{\hat{f}}_- := \int_0^\pi \hat{f}_-(t)\,\varphi(t)\,dt > -\overline{h}.$$

Then the problem (2.14) *has at least one solution.*

Proof : We will prove that condition (i) of Theorem 2.9 implies condition (i) of Theorem 2.8.

Observe first that (i) implies that there exist r and $k_+ \in L^1(0,\pi)$ such that, for a.e. $t \in [0,\pi]$ and all $x \geq r$, we have

$$f(t,x) \leq k_+(t)$$

and

$$\overline{k}_+ := \int_0^\pi k_+(t)\,\varphi(t)\,dt < -\overline{h}.$$

Moreover, as f is a L^1-Carathéodory function, we have $a_r \in L^1(0,\pi)$ such that, for a.e. $t \in [0,\pi]$ and all $x \geq 0$,

$$f(t,x) \leq |k_+(t)| + a_r(t).$$

Let ϵ be such that

$$\int_\epsilon^{\pi-\epsilon} k_+(t)\,\varphi(t)\,dt + \int_{[0,\pi]\setminus[\epsilon,\pi-\epsilon]} (|k_+(t)| + a_r(t))\,\varphi(t)\,dt \leq -\overline{h}$$

and s_+ be such that $s_+\,\varphi(t) \geq r$ for all $t \in [\epsilon, \pi - \epsilon]$, then, by taking

$$\begin{aligned} f_+(t) &= k_+(t), &&\text{if } t \in [\epsilon, \pi - \epsilon], \\ &= |k_+(t)| + a_r(t), &&\text{otherwise,} \end{aligned}$$

the condition (i) of Theorem 2.8 is fulfilled.   ∎

Remark 2.11 Let us notice that Theorem 2.9 is no more true if either $\overline{\hat{f}}_+ = \overline{h}$ or $\overline{\hat{f}}_- = \overline{h}$. We just have to observe that the problem

$$\ddot{u} + u + \exp(-u^2) = 0,$$
$$u(0) = 0,\ u(\pi) = 0,$$

has no solution. This can be proved multiplying the equation by $\sin t$ and integrating.

Remark 2.12 In Theorem 2.8 however, it is possible to have a solution even if $\overline{f}_+ = \overline{h}$ or $\overline{f}_- = \overline{h}$. We can consider for example the problem (2.14) with

$$\begin{aligned} f(t,u) &= \varphi(t), &&\text{if } u < -\varphi(t), \\ &= -u, &&\text{if } -\varphi(t) \leq u < \varphi(t), \\ &= -\varphi(t), &&\text{if } \varphi(t) \leq u \end{aligned}$$

and $h(t) = \varphi(t)$ or $h(t) = -\varphi(t)$.

Remark 2.13 In the same Theorem 2.8, the assumptions

$$u \le s_- < 0 \Rightarrow f(t, u) \ge f_-(t) \text{ and } u \ge s_+ > 0 \Rightarrow f(t, u) \le f_+(t)$$

are not enough to have the result as shown by the problem (2.14) with

$$f(t, u) = \cos u \sin t, \quad \text{if } u \in [-\pi/2, \pi/2],$$
$$= 0, \qquad\qquad \text{otherwise,}$$

and $h = 0$.

We can also prove the following result.

Theorem 2.10 Let $f : [0, \pi] \times \mathbb{R} \to \mathbb{R}$ be a L^1-Carathéodory function. Assume $f(t, .)$ is a nonincreasing function for each $t \in [0, \pi]$. Then the problem

$$\ddot{u} + u + f(t, u) = 0,$$
$$u(0) = 0, \ u(\pi) = 0,$$

has at least one solution if and only if the function $f^* : \mathbb{R} \to \mathbb{R}$ defined by

$$f^*(\xi) = \int_0^\pi f(t, \xi \sin t) \sin t \, dt$$

has a zero.

Exercise 2.6 Prove the previous result.
Hint : See J. Mawhin [89].

Problem 2.7 Prove the same kind of results for the periodic boundary value problem.

2.3 Degree Theory

The Periodic Problem

Consider first the periodic problem

$$\ddot{u} + f(t, u) = 0,$$
$$u(0) = u(2\pi), \ \dot{u}(0) = \dot{u}(2\pi), \tag{2.15}$$

which can be written under the form

$$u(t) = (Tu)(t) := \int_0^{2\pi} G(t, s)[f(s, u(s)) + u(s)] \, ds, \tag{2.16}$$

where $G(t, s)$ is the Green function corresponding to the problem (2.4). In order to give a meaning to

$$\deg(I - T, \Omega)$$

where

$$\Omega = \{u \in C([0, 2\pi]) \mid \forall t \in [0, 2\pi], \alpha(t) < u(t) < \beta(t)\}, \tag{2.17}$$

we want to define curves

$$u = \alpha(t) \quad \text{and} \quad u = \beta(t)$$

so that solution curves of (2.15) cannot be tangent to them, respectively from above or from below. The classical way to obtain such a result in the case of a continuous f and $\alpha \in C^2([0, 2\pi])$ is described in the next proposition.

Proposition 2.11 *Let f be continuous and $\alpha \in C^2([0, 2\pi])$ be such that*
(i) for all $t \in [0, 2\pi]$, $\ddot{\alpha}(t) + f(t, \alpha(t)) > 0$;
(ii) $\alpha(0) = \alpha(2\pi)$, $\dot{\alpha}(0) \geq \dot{\alpha}(2\pi)$.
If $u \in C^2([0, 2\pi])$ is a solution of (2.15) with $u \geq \alpha$ on $[0, 2\pi]$ then, for all $t \in [0, 2\pi]$, $u(t) > \alpha(t)$.

Proof : Assume, by contradiction, that

$$\min_t (u(t) - \alpha(t)) = u(t_0) - \alpha(t_0) = 0.$$

We have $\dot{u}(t_0) - \dot{\alpha}(t_0) = 0$; in case $t_0 = 0$ or 2π, this follows from assumption (ii). Hence we obtain the contradiction

$$0 \leq \ddot{u}(t_0) - \ddot{\alpha}(t_0) = -f(t_0, \alpha(t_0)) - \ddot{\alpha}(t_0) < 0. \qquad \blacksquare$$

If f is not continuous but L^p-Carathéodory, this last result does not hold anymore. In fact, even the stronger condition

$$\text{for a.e. } t \in [0, 2\pi], \quad \ddot{\alpha}(t) + f(t, \alpha(t)) \geq 1$$

does not prevent solutions $u(t)$ of (2.15) to be tangent to the curve $u = \alpha(t)$ from above. This is, for example, the case for the bounded function

$$\begin{aligned} f(t, u) &:= 1 & u \leq -1, \\ &:= \frac{u^2 + \sin t}{1 + \sin t} & -1 < u \leq \sin t, t \neq \frac{3\pi}{2} \\ &:= \sin t & \sin t < u, \end{aligned}$$

if we consider $\alpha(t) \equiv -1$ and $u(t) \equiv \sin t$. This remark motivates the following definition.

Definition 2.4 *A function $\alpha \in C_{2\pi}$ is a strict $W^{2,1}$-lower solution (or a strict lower solution) of (2.15) if it is not a solution on $[0, 2\pi]$ and for any $t_0 \in \mathbb{R}$, either $D_-\alpha(t_0) < D^+\alpha(t_0)$, or there exist an open interval I_0 and $\epsilon_0 > 0$ such that $t_0 \in I_0$, $\alpha \in W^{2,1}(I_0)$ and, for almost every $t \in I_0$, for all u with $\alpha(t) \leq u \leq \alpha(t) + \epsilon_0$, we have,*

$$\ddot{\alpha}(t) + f(t, u) \geq 0.$$

In the same way, a function $\beta \in C_{2\pi}$ is a strict $W^{2,1}$-upper solution (or a strict upper solution) of (2.15) if it is not a solution on $[0, 2\pi]$ and for any $t_0 \in \mathbb{R}$, either $D^-\beta(t_0) > D_+\beta(t_0)$, or there exist an open interval I_0 and $\epsilon_0 > 0$ such that $t_0 \in I_0$, $\beta \in W^{2,1}(I_0)$ and, for almost every $t \in I_0$, for all u with $\beta(t) - \epsilon_0 \leq u \leq \beta(t)$, we have,

$$\ddot{\beta}(t) + f(t, u) \leq 0.$$

Remark 2.14 Observe that, in the continuous case, if α satisfies the conditions of Proposition 2.11, then it is a strict $W^{2,1}$-lower solution. The converse, however, does not hold since the above definition does not require strict inequalities.

Proposition 2.12 Let f be a L^1-Carathéodory function and $\alpha \in C_{2\pi}$ be a strict $W^{2,1}$-lower solution. If $u \in W^{2,1}(0, 2\pi)$ is a solution of (2.15) with $u \geq \alpha$ on $[0, 2\pi]$ then, for all $t \in [0, 2\pi]$, $u(t) > \alpha(t)$.

Proof : As α is not a solution, $u \neq \alpha$ and there exists t^* such that $u(t^*) > \alpha(t^*)$. Assume, by contradiction, that

$$t_0 = \inf\{t > t^* \mid u(t) = \alpha(t)\}$$

exists. As $u - \alpha$ is minimum at t_0, we have $D_-\alpha(t_0) \geq D^+\alpha(t_0)$ and, from the definition of lower solution, we can assume there exist I_0, $\epsilon_0 > 0$ and $t_1 \in I_0$ with $t_1 < t_0$ such that, for every $t \in]t_1, t_0[$, $u(t) \leq \alpha(t) + \epsilon_0$, $(\dot{u} - \dot{\alpha})(t_1) < 0$ and, for almost every $t \in]t_1, t_0[$,

$$\ddot{\alpha}(t) + f(t, u(t)) \geq 0.$$

Hence, we have the contradiction

$$0 < (\dot{u} - \dot{\alpha})(t_0) - (\dot{u} - \dot{\alpha})(t_1) = -\int_{t_1}^{t_0}[f(t, u(t)) + \ddot{\alpha}(t)]\, dt \leq 0. \qquad \blacksquare$$

Now we can prove the key result of this section.

Theorem 2.13 Let α and β be strict $W^{2,1}$-lower and upper solutions of the problem (2.15) such that $\alpha < \beta$ on $[0, 2\pi]$. Define E from

$$E := \{(t, u) \in [0, 2\pi] \times \mathbb{R} \mid \alpha(t) \leq u \leq \beta(t)\}. \tag{2.18}$$

Assume $f : E \to \mathbb{R}$ is a L^1-Carathéodory function.
    Then

$$\deg(I - T, \Omega) = 1, \tag{2.19}$$

where T and Ω are defined by (2.16) and (2.17).
    Moreover, the problem (2.15) has at least one solution $u \in W^{2,1}(0, 2\pi)$ such that, for all $t \in [0, 2\pi]$,

$$\alpha(t) < u(t) < \beta(t).$$

Proof : Let us consider the modified problem

$$\ddot{u} - u + f(t, \gamma(t, u)) + \gamma(t, u) = 0,$$
$$u(0) = u(2\pi), \quad \dot{u}(0) = \dot{u}(2\pi),$$

(2.20)

where $\gamma : [0, 2\pi] \times \mathbb{R} \to \mathbb{R}$ is defined by (2.3).
    Define the operator

$$\overline{T} : C([0, 2\pi]) \to C([0, 2\pi]), u \to \int_0^{2\pi} G(t, s)[f(s, \gamma(s, u(s))) + \gamma(s, u(s))] \, ds,$$

where $G(t, s)$ is the Green function corresponding to the problem (2.4). It is clear that $\mathrm{Im}\overline{T}$ is bounded and, for $\overline{R} > 0$ large enough and any $\lambda \in [0, 1]$,

$$\deg(I - \overline{T}, B(0, \overline{R})) = \deg(I - \lambda\overline{T}, B(0, \overline{R})) = \deg(I, B(0, \overline{R})) = 1.$$

It follows from the proof of Theorem 2.2 and Proposition 2.12 that every solution u of (2.20) satisfies $\alpha(t) < u(t) < \beta(t)$ on $[0, 2\pi]$. This proves that such a solution is in Ω. As T and $\overline{T}$ coincide on $\overline{\Omega}$, we obtain, by the excision property of the degree,

$$\deg(I - T, \Omega) = \deg(I - \overline{T}, \Omega) = \deg(I - \overline{T}, B(0, \overline{R})) = 1$$

which proves the result.   ∎

Forced Pendulum Type Equation

In Example 2.2, we have proved that if $h \in C([0, 2\pi])$ is such that $\|h\|_\infty < 1$, the problem

$$\ddot{u} + \sin u = h(t),$$
$$u(0) = u(2\pi), \quad \dot{u}(0) = \dot{u}(2\pi),$$

(2.21)

has at least one solution. With Theorem 2.13, it is now easy to complement this result.

Proposition 2.14 *Assume that $h \in C([0, 2\pi])$ and $\|h\|_\infty < 1$, then the problem (2.21) has at least two solutions such that, for all $k \in \mathbb{Z}$ there exists $t \in [0, 2\pi]$ with $u_1(t) \neq u_2(t) + 2k\pi$.*

Proof : Observe that $\alpha_1(t) := \pi/2$ and $\alpha_2(t) := \pi/2 + 2\pi$ are strict lower solutions and $\beta_1(t) := 3\pi/2$ and $\beta_2(t) := 3\pi/2 + 2\pi$ are strict upper solutions of (2.21).
    Now we apply Theorem 2.13 successively with

$$\Omega_1 = \{u \in C([0, 2\pi]) \mid \forall t \in [0, 2\pi], \alpha_1(t) < u(t) < \beta_1(t)\},$$
$$\Omega_2 = \{u \in C([0, 2\pi]) \mid \forall t \in [0, 2\pi], \alpha_2(t) < u(t) < \beta_2(t)\}$$

and

$$\Omega = \{u \in C([0, 2\pi]) \mid \forall t \in [0, 2\pi], \alpha_1(t) < u(t) < \beta_2(t)\}.$$

As

$$\deg(I - T, \Omega_1) = \deg(I - T, \Omega_2) = 1,$$

we have two solutions u_1, u_2 with $\pi/2 < u_1 < 3\pi/2$ and $5\pi/2 < u_2 < 7\pi/2$. Moreover, we have

$$1 = \deg(I - T, \Omega)$$
$$= \deg(I - T, \Omega_1) + \deg(I - T, \Omega_2) + \deg(I - T, \Omega \setminus (\overline{\Omega}_1 \cup \overline{\Omega}_2)),$$

which implies
$$\deg(I - T, \Omega \setminus (\overline{\Omega}_1 \cup \overline{\Omega}_2)) = -1$$

and hence we have the existence of a third solution $u_3 \in \Omega \setminus (\overline{\Omega}_1 \cup \overline{\Omega}_2)$. The solution u_2 might be of the form $u_2(t) = u_1(t) + 2\pi$ but u_3 cannot equal u_1 modulo 2π, which proves the result. ∎

This proposition is an example of the three solutions Theorem (see H. Amann [9]). Forced pendulum have been extensively studied using lower and upper solutions as well as other methods (see [89], [96], [87], [92], [73] for more informations and more results).

The Dirichlet Problem

Consider the boundary value problem

$$\ddot{u} + f(t, u) = 0,$$
$$u(0) = 0, \ u(\pi) = 0,$$
(2.22)

where f is a L^1-Carathéodory function.

Again, we need first to define the notion of strict lower and upper solutions for (2.22). Here the situation differs however from the periodic case because we want to accept the possibility for upper and lower solutions to satisfy the boundary conditions which implies that at boundary points, the solution can equal upper and lower solutions.

Definition 2.5 *A function $\alpha \in C([0, \pi])$ is a strict $W^{2,1}$-lower solution (or a strict lower solution) of (2.22) if it is not a solution of (2.22), $\alpha(0) \leq 0$, $\alpha(\pi) \leq 0$ and for any $t_0 \in [0, \pi]$, one of the following is satisfied:*
(i) $t_0 \in \{0, \pi\}$ and $\alpha(t_0) < 0$;
(ii) $t_0 \in]0, \pi[$ and $D_-\alpha(t_0) < D^+\alpha(t_0)$;
(iii) there exist an interval $I_0 \subset [0, \pi]$ and $\epsilon > 0$ such that $t_0 \in \text{int} I_0$ or $t_0 \in I_0 \cap \{0, \pi\}$, $\alpha \in W^{2,1}(I_0)$ and for almost every $t \in I_0$, for all $u \in [\alpha(t), \alpha(t) + \epsilon \sin t]$ we have

$$\ddot{\alpha}(t) + f(t, u) \geq 0.$$

In a similar way, a function $\beta \in C([0, \pi])$ is a strict $W^{2,1}$-upper solution (or a strict upper solution) of (2.22) if it is not a solution of (2.22), $\beta(0) \geq 0$, $\beta(\pi) \geq 0$ and for any $t_0 \in [0, \pi]$, one of the following is satisfied:
(i) $t_0 \in \{0, \pi\}$ and $\beta(t_0) > 0$;
(ii) $t_0 \in]0, \pi[$ and $D^-\beta(t_0) > D_+\beta(t_0)$;

(iii) there exist an interval $I_0 \subset [0, \pi]$ and $\epsilon > 0$ such that $t_0 \in \text{int} I_0$ or $t_0 \in I_0 \cap \{0, \pi\}$, $\beta \in W^{2,1}(I_0)$ and for almost every $t \in I_0$, for all $u \in [\beta(t) - \epsilon \sin t, \beta(t)]$ we have

$$\ddot{\beta}(t) + f(t, u) \le 0.$$

In this definition the curve $u = \beta(t)$ can have angles, provided they are downward. Also, $\beta(0)$ and $\beta(\pi)$ can be zero. In this case, condition (iii) imposes some second order condition near $t = 0$ or $t = \pi$ but restricted to some angular region below the curve $u = \beta(t)$. Notice at last that (i) can be interpreted as a zero order condition and (ii) as a first order one in order to prevent solutions to be tangent to α or β.

Remark 2.15 If f is continuous, a function $\beta \in C^2(]0, \pi[)$ such that $\beta(0) > 0$, $\beta(\pi) > 0$ and

$$\forall t \in]0, \pi[, \quad \ddot{\beta}(t) + f(t, \beta(t)) < 0$$

is a strict $W^{2,1}$-upper solution.

Also, if the function $\beta \in C^2([0, \pi])$ is such that $\beta(0) \ge 0$, $\beta(\pi) \ge 0$ and

$$\forall t \in [0, \pi], \quad \ddot{\beta}(t) + f(t, \beta(t)) < 0,$$

it is a strict $W^{2,1}$-upper solution.

As in the periodic case, the converse does not hold as strict $W^{2,1}$-upper solutions do not necessarily satisfy $\ddot{\beta}(t) + f(t, \beta(t)) < 0$.

Lower solutions must be smaller than upper ones. To make this precise in case these functions coincide at end points, we introduce the following notation.

Definition 2.6 *Let $x, y \in C([0, \pi])$. We write $x \prec y$ if there exists $\epsilon > 0$ such that, for any $t \in [0, \pi]$,*

$$y(t) - x(t) \ge \epsilon \sin t.$$

Observe also that, even in the continuous case, it is not enough to have

$$\ddot{\beta}(t) + f(t, \beta(t)) < 0 \text{ on }]0, \pi[, \quad \beta(0) \ge 0, \quad \beta(\pi) \ge 0$$

to be sure that every solution u of (2.22) with $u \le \beta$ is such that $u \prec \beta$. To see that, we just have to consider the functions

$$
\begin{aligned}
f(t, u) &= 0 & &\text{if } u \le 0, \\
&= -7u/t^2 & &\text{if } 0 < u \le t^3, \\
&= -7t & &\text{if } u > t^3,
\end{aligned}
$$

$u(t) = 0$ and $\beta(t) = t^3$.

The main result of this section is a multiplicity result.

Theorem 2.15 *Assume f is a L^1-Carathéodory function. Let α, β be strict $W^{2,1}$-lower and upper solutions of (2.22) such that $\alpha \prec \beta$. Assume moreover, there exists $r > 0$, such that for all $s < 0$ and all solutions u of*

$$\ddot{u} + f(t, u) = s,$$
$$u(0) = 0, \ u(\pi) = 0,$$
(2.23)

we have

$$\|u\|_\infty < r.$$

Then the problem (2.22) has at least two solutions u_1, $u_2 \in W^{2,1}(0, \pi)$.

Proof : Observe first that the problem (2.22) is equivalent to the fixed point equation

$$u = Tu := \int_0^\pi G(t, s) f(s, u(s)) \, ds,$$

where $G(t, s)$ is the Green function of (2.11). The operator

$$T : \mathcal{C}_0^1([0, \pi]) \to \mathcal{C}_0^1([0, \pi])$$

is completely continuous. Let $r > \max\{\|\alpha\|_\infty, \|\beta\|_\infty\}$ and define $R > 0$ such that every solution u of (2.22) with $\|u\|_\infty < r$ satisfies $\|\dot{u}\|_\infty < R$.

From the argument in Theorem 2.13, we prove that, if R is large enough,

$$\deg(I - T, \Omega) = 1,$$

where

$$\Omega := \{u \in \mathcal{C}_0^1([0, \pi]) \mid \alpha \prec u \prec \beta, \|\dot{u}\|_\infty < R\}.$$

In this way, we have the existence of a first solution u_1 of (2.22).

To prove the existence of the second solution, let us first compute the degree

$$\deg(I - T, \Omega_1),$$

where

$$\Omega_1 := \{u \in \mathcal{C}_0^1([0, \pi]) \mid \|u\|_\infty < r, \ \|\dot{u}\|_\infty < R\}.$$

We have assumed that, for all $s < 0$, the solutions u of

$$\ddot{u} + f(t, u) = s,$$
$$u(0) = 0, \ u(\pi) = 0,$$

i.e. the solutions of

$$u = Tu - s\,h,$$
(2.24)

with $h(t) := \frac{t(\pi - t)}{2}$ are such that

$$\|u\|_\infty < r.$$

We can assume this is true also for $s = 0$ since otherwise the existence of a second solution is proved. As T maps bounded sets into bounded sets, there exists $r_1 > 0$ such that, for all $u \in \overline{\Omega}_1$,

$$\|u - Tu\|_\infty < r_1.$$

Hence, if s is large enough, the problem (2.24) has no solution and

$$\deg(I - T, \Omega_1) = \deg(I - T + sh, \Omega_1) = 0.$$

As $\Omega \subset \Omega_1$, we obtain the existence of the second solution by the excision property of the degree.   ∎

Remark 2.16 In this theorem, it is essential to define T on $C_0^1([0, \pi])$, since

$$\Omega^* := \{u \in C_0([0, \pi]) \mid \alpha \prec u \prec \beta\}$$

is not open in the C-topology.

Remark 2.17 Notice that the condition on the solutions of (2.23) is an a priori bound on some upper solutions.

The reader can find an application of this result in Section 3.2.

As in Section 2.2, we can consider the case where f satisfies a Carathéodory condition and there exists a function $h \in \mathcal{A}$ such that,

$$|f(t, u)| \leq h(t).$$

Recall that
$$\mathcal{A} = \{h \in L^1_{loc}(]0, \pi[) \mid s(\pi - s)h(s) \in L^1(0, \pi)\}.$$

As the solution is now in $W^{2,\mathcal{A}}(0, \pi) \not\subset C^1([0, \pi])$, we can no more define the operator T on $C_0^1([0, \pi])$. Nevertheless, at least if $\beta(0) > 0$ and $\beta(\pi) > 0$, we have the following result.

Theorem 2.16 *Let α be a $W^{2,1}$-lower solution of the problem (2.22) and β a strict $W^{2,1}$-upper solution of (2.22) such that $\alpha \leq \beta$, $\beta(0) > 0$ and $\beta(\pi) > 0$. Assume $f : [0, \pi] \times \mathbb{R} \to \mathbb{R}$ satisfies a Carathéodory condition and, for all $R > 0$ there exists $h \in \mathcal{A}$ such that, for a.e. $t \in [0, \pi]$ and for all $u \in [\alpha(t), R]$,*

$$|f(t, u)| \leq h(t).$$

If moreover, there exists $r > 0$ such that, for all $s < 0$ and, for all solutions u of

$$\ddot{u} + f(t, u) = s,$$
$$u(0) = 0, \ u(\pi) = 0,$$

with $u \geq \alpha$, we have

$$u(t) < r,$$

then the problem (2.22) has at least two solutions $u_1, u_2 \in W^{2,\mathcal{A}}(0, \pi)$ such that, for all $t \in [0, \pi]$,

$$\alpha(t) \leq u_1(t) < \beta(t), \quad \alpha(t) \leq u_2(t)$$

and, there exists $t_0 \in [0, \pi]$ with

$$u_2(t_0) > \beta(t_0) > u_1(t_0).$$

Exercise 2.8 Prove the previous theorem.
Hint : See C. De Coster, M.R. Grossinho and P. Habets [35].

Application can be found in Section 2.2.

Problem 2.9 Extend Theorem 2.15 to the separated boundary value problem

$$\ddot{u} = f(t, u),$$
$$a_1 u(a) - a_2 \dot{u}(a) = A,$$
$$b_1 u(b) + b_2 \dot{u}(b) = B,$$

in case $A, B \in \mathbb{R}$, $a_1, b_1 \in \mathbb{R}$, $a_2, b_2 \in \mathbb{R}^+$, $a_1^2 + a_2^2 > 0$ and $b_1^2 + b_2^2 > 0$.
Hint : See [33], [15].

2.4   Variational Methods

Another approach in working with lower and upper solutions is to relate them with variational methods. More precisely, when we consider the problem

$$\ddot{u} + f(t, u) = 0,$$
$$u(0) = 0, \quad u(\pi) = 0, \tag{2.25}$$

where f is a L^1-Carathéodory function, it is well known that the related functional

$$\phi : H_0^1(0, \pi) \to \mathbb{R}, u \to \int_0^\pi [\frac{\dot{u}^2(t)}{2} - F(t, u(t))] \, dt, \tag{2.26}$$

with $F(t, s) = \int_0^s f(t, x) \, dx$, is of class C^1 and its critical points are the solutions of (2.25).

A first result of this approach is the following.

Theorem 2.17 *Let α and β be $W^{2,1}$-lower and upper solutions of (2.25) with $\alpha \leq \beta$ on $[0, \pi]$. Assume f satisfies Carathéodory conditions and there exists $h \in L^1(0, \pi)$ such that, for a.e. $t \in [0, \pi]$ and all $u \in [\alpha(t), \beta(t)]$, $|f(t, u)| \leq h(t)$. Then the problem (2.25) has a solution u with $\alpha \leq u \leq \beta$ on $[0, \pi]$ and*

$$\phi(u) = \min_{\substack{v \in H_0^1(0,\pi) \\ \alpha \leq v \leq \beta}} \phi(v).$$

Proof : Let us consider the modified problem

$$\ddot{u} + f(t, \gamma(t, u)) = 0,$$
$$u(0) = 0, \ u(\pi) = 0,$$
(2.27)

where γ is defined as in (2.3). Define the functional

$$\psi : H_0^1(0, \pi) \to \mathbb{R}, u \to \int_0^\pi [\frac{\dot{u}^2(t)}{2} - \overline{F}(t, u(t))]dt,$$

where $\overline{F}(t, s) = \int_0^s f(t, \gamma(t, x)) \, dx$. It is easy to verify that ψ is of class $\mathcal{C}^1$ and its critical points are precisely the solutions of (2.27). Moreover ψ is weakly lower semicontinuous and coercive. Hence, ψ has a global minimum u which is a solution of (2.27). As in Section 2.2, we prove that u satisfies $\alpha \leq u \leq \beta$, is a solution of (2.25) and hence is such that

$$\phi(u) = \min_{\substack{v \in H_0^1(0, \pi) \\ \alpha \leq v \leq \beta}} \phi(v).$$ ∎

Let us now consider the problem

$$\ddot{u} + \mu(t)g(u) + h(t) = 0,$$
$$u(0) = 0, \ u(\pi) = 0.$$
(2.28)

The following result is due to P. Omari and F. Zanolin [104] who consider an elliptic p-Laplacian. For the details of the proof, we refer to [104].

Theorem 2.18 *Let $\mu, h \in L^\infty(0, \pi)$ with $\mu_0 = \mathrm{essinf}\, \mu(t) > 0$ and $g : \mathbb{R} \to \mathbb{R}$ be a continuous function. Let us denote $G(u) = \int_0^u g(s) \, ds$. Assume that*

$$-\infty < \liminf_{s \to \pm\infty} \frac{G(s)}{s^2} \leq 0 \quad and \quad \limsup_{s \to \pm\infty} \frac{G(s)}{s^2} = +\infty.$$

Then the problem (2.28) has two infinite sequences $(u_n)_n$ and $(v_n)_n$ of solutions satisfying

$$\ldots \leq v_{n+1} \leq v_n \leq \ldots \leq v_1 \leq u_1 \leq \ldots \leq u_n \leq u_{n+1} \leq \ldots$$

and

$$\lim_{n \to \infty} (\max_t u_n(t)) = +\infty, \quad \lim_{n \to \infty} (\min_t v_n(t)) = -\infty.$$

Proof : Step 1 – Claim : For every $M \geq 0$ there exists β an upper solution of (2.28) with $\beta(t) \geq M$ on $[0, \pi]$.

First observe that, if g is unbounded from below on $[0, +\infty[$, we have a sequence of constant upper solutions $\beta_n \to +\infty$. Therefore, we can assume there exists $K \geq 0$ such that $g(s) \geq -K$ for $s \geq 0$.

Given $M > 0$, let us choose d large enough so that

$$\|\mu\|_\infty \left(\frac{G(d)}{d^2} + \frac{K}{d} \right) + \frac{\|h\|_\infty}{d} \leq \frac{1}{8\pi^2} \quad \text{and} \quad d > 2M$$

and define β to be the solution of the Cauchy problem

$$\ddot{u} + \|\mu\|_\infty (g(u) + K) + \|h\|_\infty = 0,$$
$$u(0) = d, \ \dot{u}(0) = 0. \tag{2.29}$$

Assume there exists $t_0 \in \,]0, \pi]$ such that $\beta(t) > M$ on $[0, t_0[$ and $\beta(t_0) = M$. On $[0, t_0]$, $\dot{\beta}(t) \leq 0$, $\|\mu\|_\infty (G(\beta(t)) + K\beta(t)) + \|h\|_\infty \beta(t) \geq 0$ and from the conservation of energy for (2.29), we have

$$\frac{\dot{\beta}^2(t)}{2} \leq \frac{\dot{\beta}^2(t)}{2} + \|\mu\|_\infty (G(\beta(t)) + K\beta(t)) + \|h\|_\infty \beta(t)$$

$$= \|\mu\|_\infty (G(d) + Kd) + \|h\|_\infty d \leq \frac{d^2}{8\pi^2},$$

i.e.

$$0 \leq -\dot{\beta}(t) \leq \frac{d}{2\pi}.$$

It follows that for any $t \in [0, t_0]$,

$$d - \beta(t) \leq \frac{d}{2\pi}\pi = \frac{d}{2},$$

which leads to the contradiction $\beta(t_0) \geq \frac{d}{2} > M$. Hence the claim follows.

In a similar way, we prove

Claim – For every $M \geq 0$ there exists a lower solution α of (2.28) such that $\alpha \leq -M$ on $[0, \pi]$.

Step 2 – By simple but not obvious estimates we can show that there exists a sequence of positive real numbers $(s_n)_n$ with $s_n \to +\infty$, a sequence of negative real numbers $(t_n)_n$ with $t_n \to -\infty$ and $z \succ 0$ such that $\phi(s_n z) \to -\infty$ and $\phi(t_n z) \to -\infty$.

Step 3 – Conclusion. By Step 1, we have α_1, β_1 lower and upper solutions of (2.28) with $\alpha_1 \leq \beta_1$. Hence, we have a solution u_1 of (2.28) such that $\alpha_1 \leq u_1 \leq \beta_1$ and

$$\phi(u_1) = \min_{\substack{v \in H_0^1(0,\pi) \\ \alpha_1 \leq v \leq \beta_1}} \phi(v).$$

From Step 2, we have s_1 such that $u_1 \leq s_1 z$ and $\phi(u_1) > \phi(s_1 z)$. Moreover, Step 1 provides the existence of an upper solution β_2 with $u_1 \leq s_1 z \leq \beta_2$. Hence, by Theorem 2.17, we have a solution u_2 of (2.28) satisfying

$$u_1 \leq u_2 \leq \beta_2$$

and

$$\phi(u_2) = \min_{\substack{v \in H_0^1(0,\pi) \\ u_1 \leq v \leq \beta_2}} \phi(v) \leq \phi(s_1 z) < \phi(u_1).$$

It follows that $u_2 \neq u_1$. Iterating this argument and reproducing it in the negative part, we prove the result. ∎

2.5   Monotone Iterative Methods

In this section, we consider the problem

$$\ddot{u} + f(t, u) = 0,$$
$$u(0) = u(2\pi), \ \dot{u}(0) = \dot{u}(2\pi),$$
(2.30)

where f is a continuous function.

The monotone iterative method consists in generating sequences whose first term are respectively lower and upper solutions and that monotonically converge to the minimal and maximal solutions of (2.30).

Definition 2.7 *We say that u is a* maximal solution *of (2.30) in $[\alpha, \beta]$ (resp: a* minimal solution *of (2.30) in $[\alpha, \beta]$) if it is a solution with $\alpha \le u \le \beta$ on $[0, 2\pi]$ and every solution v of (2.30) with $\alpha \le v \le \beta$ is such that $v \le u$ on $[0, 2\pi]$ (resp: $u \le v$ on $[0, 2\pi]$).*

The key ingredient of this method is the following maximum principle.

Lemma 2.19 *Let $M > 0$ and assume that $u \in C^2([0, 2\pi])$ is such that*

$$\ddot{u} - Mu \le 0 \ on \ [0, 2\pi],$$
$$u(0) = u(2\pi), \ \dot{u}(0) \le \dot{u}(2\pi).$$

Then $u(t) \ge 0$ on $[0, 2\pi]$.

Proof : Assume by contradiction there exists $t_1 \in [0, 2\pi]$ such that $\min_t u = u(t_1) < 0$. Thanks to the boundary conditions, it is easy to see that $\dot{u}(t_1) = 0$ and we have the contradiction

$$0 \le \ddot{u}(t_1) \le Mu(t_1) < 0. \qquad \blacksquare$$

Observe that u is a solution of (2.30) if and only if

$$u = Tu := \int_0^{2\pi} G(t, s)[f(s, u(s)) + Mu(s)] \, ds,$$

where $G(t, s)$ is the Green function of

$$\ddot{u} - Mu + h(t) = 0,$$
$$u(0) = u(2\pi), \ \dot{u}(0) = \dot{u}(2\pi).$$

Now we can state the main result of this section.

Theorem 2.20 *Let $f : [0, 2\pi] \times \mathbb{R} \to \mathbb{R}$ be continuous and let α and β be lower and upper solutions of (2.30) with $\alpha \le \beta$. Assume there exists $M > 0$ such that, for all $t \in [0, 2\pi]$ and any u, v with $\alpha(t) \le u \le v \le \beta(t)$, we have*

$$f(t, v) - f(t, u) \ge -M(v - u).$$
(2.31)

Then the sequences $(\alpha_n)_n$ and $(\beta_n)_n$ defined by

$$\alpha_0 = \alpha, \ \alpha_{n+1} = T\alpha_n,$$
$$\beta_0 = \beta, \ \beta_{n+1} = T\beta_n, \qquad (2.32)$$

converge uniformly to the minimal and the maximal solution of (2.30) in $[\alpha, \beta]$.

Proof : Let us prove first that T is monotone on $[\alpha, \beta]$. Let u and v be such that $\alpha \leq u \leq v \leq \beta$. We just have to observe that $u_1 = Tu$ and $v_1 = Tv$ are such that

$$(\ddot{v}_1 - \ddot{u}_1) - M(v_1 - u_1) = -f(t, v(t)) + f(t, u(t)) - M(v(t) - u(t)) \leq 0,$$
$$(v_1 - u_1)(0) = (v_1 - u_1)(2\pi), \ (\dot{v}_1 - \dot{u}_1)(0) = (\dot{v}_1 - \dot{u}_1)(2\pi).$$

By Lemma 2.19 we have $Tv - Tu = v_1 - u_1 \geq 0$.

In the same way, we prove that $\alpha \leq T\alpha$ and $\beta \geq T\beta$ and

$$\alpha = \alpha_0 \leq \alpha_1 \leq \ldots \leq \alpha_n \leq \ldots \leq \beta_n \leq \ldots \leq \beta_1 \leq \beta_0 = \beta.$$

As T is completely continuous and the sequences $(\alpha_n)_n$ and $(\beta_n)_n$ are bounded, we have for some subsequences

$$\alpha_{n_k} \to \underline{u}, \ \beta_{n_k} \to \overline{u}$$

uniformly on $[0, 2\pi]$. Next, by the monotonicity of $(\alpha_n)_n$ and $(\beta_n)_n$ we prove $\alpha_n \to \underline{u}$, $\beta_n \to \overline{u}$. Going to the limit in (2.32), $\underline{u}$ and $\overline{u}$ are solutions of (2.30).

To show that every solution $u \in [\alpha, \beta]$ satisfies $u \in [\underline{u}, \overline{u}]$, we just have to observe that, as $\alpha \leq u \leq \beta$ and T is monotone, we have

$$\alpha_n = T^n \alpha \leq T^n u = u \leq T^n \beta = \beta_n$$

and the result follows going to the limit.   ∎

Remark 2.18 We can have the existence of the maximal and the minimal solution even without the condition (2.31), but we loose the converging sequence (see K. Ako [5], [3] and also K. Schmitt [110]).

Problem 2.10 Extend this result to the L^1-Carathéodory case.

Problem 2.11 Extend Theorem 2.20 to the case where (2.31) is replaced by

$$f(t, v) - f(t, u) \geq -M(v - u)^r,$$

with $0 < r \leq 1$.
Hint : See H. Amann [6] or J.I. Diaz [44].

3   Systems with singularities

Boundary value problems

$$\ddot{u} + f(t, u) = 0,$$
$$u(0) = 0, \ u(\pi) = 0,$$

(3.1)

where the function f is singular both at the end points $t = 0, t = \pi$ and for $u = 0$, have been used in several applied mathematics problem [14]. A typical example is the generalized Emden-Fowler equation

$$\ddot{u} + \frac{f(t)}{u^\sigma} = 0,$$
$$u(0) = 0, \ u(\pi) = 0.$$

(3.2)

In case f is singular at end points, we know from Theorem 2.6 that a natural setting is to look for solutions $u \in W^{2,\mathcal{A}}(0, \pi)$. For some singularities at $u = 0$, solutions are still in $W^{2,\mathcal{A}}(0, \pi)$. This is the case for the problem

$$\ddot{u} + \frac{1}{(\pi - t)t} \frac{1}{u} = 0,$$
$$u(0) = 0, \ u(\pi) = 0,$$

(3.3)

whose solution

$$u(t) = \frac{2}{\pi} t^{\frac{1}{2}} (\pi - t)^{\frac{1}{2}}$$

is in this space. Notice that $f(t, u) = \frac{1}{(\pi-t)t} \frac{1}{u}$ is only $\mathcal{A}$-Carathéodory on sets of the form $[0, \pi] \times [\epsilon, +\infty[$, with $\epsilon > 0$ and not on $[0, \pi] \times]0, +\infty[$. As the boundary conditions impose the solution to go into the singularity $u = 0$, there is no hope to obtain a positive solution from a simple application of Theorem 2.6. In this section we shall consider continuous functions f such that the solutions are in $\mathcal{C}([0, \pi], \mathbb{R}^+) \cap \mathcal{C}^2(]0, \pi[, \mathbb{R}_0^+)$.

3.1   Existence of a positive solution

The following theorem, which can be found in P. Habets and F. Zanolin [64], considers the boundary value problem (3.1). Recall we defined $\mathcal{A}$ to be the set of functions $h : [0, \pi] \to \mathbb{R}^+$ measurable such that

$$\int_0^\pi s(\pi - s)h(s) \, ds < +\infty.$$

Theorem 3.1 *Assume :*
(i)   the function $f :]0, \pi[\times \mathbb{R}_0^+ \to \mathbb{R}$ is continuous;
(ii) there exists $k > 1$ and for any compact set $K \subset]0, \pi[$, there is $\epsilon > 0$ such that

$$f(t, u) \geq k^2 u, \ \text{for all } t \in K, \ u \in]0, \epsilon];$$

(iii)for some $M > 0$ and $\gamma < 1$, there is $h \in A \cap C(]0, \pi[)$ such that

$$f(t, u) \leq \gamma^2 u + h(t), \quad \text{for all } t \in]0, \pi[, \ u \in [M, +\infty[;$$

(iv) for any compact set $K \subset]0, +\infty[$, there is $h_K \in A$ such that

$$|f(t, u)| \leq h_K(t), \quad \text{for all } t \in]0, \pi[, \ u \in K.$$

Then the problem (3.1) has at least one solution

$$u \in C([0, \pi], \mathbb{R}^+) \cap C^2(]0, \pi[, \mathbb{R}_0^+).$$

Remark 3.1 Assumption (ii) is equivalent to assume there exist $k > 1$ and a function $a_1 \in C_0^2([0, \pi], \mathbb{R}^+)$ such that:
(a) $t \in]0, \pi[$ implies $a_1(t) > 0$;
(b) $f(t, u) \geq k^2 u$, for all $t \in]0, \pi[, \ 0 < u \leq a_1(t)$;
(c) $\ddot{a}_1(t) > 0$, for all $t \in [0, \pi/3] \cup [2\pi/3, \pi]$.

Proof of Theorem 3.1 : Step 1 – Construction of lower solutions. Consider k_2 such that $1 < k_2 < \min(k, 2)$ and the function

$$\alpha_2(t) = A_2 \cos k_2 \left(t - \frac{\pi}{2}\right),$$

where A_2 is chosen small enough so that :

$$f(t, u) \geq k^2 u, \quad \text{for all } t \in \left]\left(1 - \frac{1}{k_2}\right)\frac{\pi}{2}, \left(1 + \frac{1}{k_2}\right)\frac{\pi}{2}\right[, \ 0 < u \leq \alpha_2(t).$$

Next, we choose a_1 from the remark and let

$$\alpha_1(t) = A_1 a_1(t),$$

where $A_1 \in]0, 1]$ is small enough so that for some points $t_1 \in]0, \frac{\pi}{3}[, \ t_2 \in]\frac{2\pi}{3}, \pi[$, one has :
(a-1)$\alpha_1(t) \geq \alpha_2(t)$, for all $t \in [0, t_1] \cup [t_2, \pi]$;
(b-1)$\alpha_2(t) \geq \alpha_1(t)$, for all $t \in [t_1, t_2]$.
   Notice that for any $h :]0, \pi[\times \mathbb{R}_0^+ \to \mathbb{R}$ such that :

$$h(t, u) \geq f(t, u), \quad \text{for all } (t, u) \in]0, \pi[\times \mathbb{R}_0^+,$$

one has :
(a-2)$\ddot{\alpha}_1(t) + h(t, \alpha_1(t)) \geq \ddot{a}_1(t) + k^2 a_1(t) > 0, \quad$ for all $t \in [0, t_1] \cup [t_2, \pi]$;
(b-2)$\ddot{\alpha}_2(t) + h(t, \alpha_2(t)) \geq -k_2^2 \alpha_2(t) + k^2 \alpha_2(t) > 0, \quad$ for all $t \in [t_1, t_2]$.

Step 2 – Approximation problems. We define for each $n \in \mathbb{N}, n \geq 1$,

$$\eta_n(t) = \max\{\frac{1}{2^{n+1}}, \min(t, \pi - \frac{1}{2^{n+1}})\}, \ t \in]0, \pi[$$

and set

$$\tilde{f}_n(t, u) = \max\{f(\eta_n(t), u), f(t, u)\}.$$

We have that, for each index n, $\tilde{f}_n :]0, \pi[\times\mathbb{R}_0^+ \to \mathbb{R}$ is continuous and

$$\tilde{f}_n(t, u) = f(t, u), \ \text{for all } (t, u) \in K_n \times \mathbb{R}_0^+,$$

where

$$K_n = [\frac{1}{2^{n+1}}, \pi - \frac{1}{2^{n+1}}].$$

Hence, the sequence of functions $\{\tilde{f}_n\}$ converges to f uniformly on any set $K \times \mathbb{R}_0^+$, where K is an arbitrary compact subset of $]0, \pi[$.

Next we define

$$f_n(t, u) = \min\{\tilde{f}_1(t, u), \cdots, \tilde{f}_n(t, u)\}.$$

Each of the functions f_i is a continuous function defined on $]0, \pi[\times\mathbb{R}_0^+$; moreover

$$f_1(t, u) \geq f_2(t, u) \geq \cdots \geq f_n(t, u) \geq f_{n+1}(t, u) \geq \cdots \geq f(t, u)$$

and the sequence (f_n) converges to f uniformly on the compact subsets of $]0, \pi[\times\mathbb{R}_0^+$ since

$$f_n(t, u) = f(t, u), \ \text{for all } t \in K_n, \ u \in \mathbb{R}_0^+.$$

Define now a decreasing sequence $(\epsilon_n) \subset \mathbb{R}_0^+$ such that

$$\lim_{n\to\infty} \epsilon_n = 0,$$

$$f(t, u) \geq k^2 u, \ \text{for all } t \in K_n, \ u \in]0, \epsilon_n],$$

and consider the sequence of approximation problems:

$$\ddot{u} + f_n(t, u) = 0,$$
$$u(0) = \epsilon_n, \ u(\pi) = \epsilon_n. \tag{P_n}$$

Step 3 – A lower solution of (P_n). It is clear that for any $c \in]0, \epsilon_n]$

$$\tilde{f}_n(t, c) \geq f(\eta_n(t), c) \geq k^2 c > 0.$$

As the sequence (ϵ_n) is decreasing, we also have

$$f_n(t, \epsilon_n) = \min_{1 \leq k \leq n} \tilde{f}_k(t, \epsilon_n) \geq k^2 \epsilon_n > 0.$$

It follows that $\alpha_3(t) := \epsilon_n$ is such that

$$\ddot{\alpha}_3(t) + f_n(t, \alpha_3(t)) = f_n(t, \epsilon_n) > 0$$

and $\alpha = \max(\alpha_1(t), \alpha_2(t), \epsilon_1)$ is a lower solution of (P_n).

Step 4 – Existence of a solution u_1 of (P_1) such that

$$\max(\alpha_1(t), \alpha_2(t), \epsilon_1) \leq u_1(t).$$

From assumption (iii), we can find $M \geq \max(\alpha_1(t), \alpha_2(t), \epsilon_1)$ and $h \in \mathcal{A}$ such that, for all $t \in]0, \pi[$, $u \in [M, +\infty[$,

$$f(t, u) \leq \gamma^2 u + h(t).$$

Also, one has

$$f(\eta_1(t), u) \leq \gamma^2 u + h(\eta_1(t)) \leq \gamma^2 u + R,$$

where $R > 0$ is a suitable constant. Hence, we can write, for such t and u,

$$f_1(t, u) = \max\{f(\eta_1(t), u), f(t, u)\} \leq \gamma^2 u + h(t) + R.$$

Choose β such that

$$\ddot{\beta} + \gamma^2 \beta + h(t) + R = 0,$$
$$\beta(0) = M, \ \beta(\pi) = M,$$

i.e.

$$\beta = M + \int_0^\pi G(t, s)[h(s) + R + \gamma^2 M] \, ds \geq M,$$

where $G(t, s)$ is the Green function of the problem

$$\ddot{u} + \gamma^2 u + g(t) = 0,$$
$$u(0) = 0, \ u(\pi) = 0.$$

Notice that β is well defined and bounded since $h \in \mathcal{A}$. It is easy to see now that

$$\ddot{\beta} + f_1(t, \beta) \leq \ddot{\beta} + \gamma^2 \beta + h(t) + R = 0.$$

By Theorem 2.6, we know that there is a solution u_1 of (P_1) such that

$$\max(\alpha_1(t), \alpha_2(t), \epsilon_1) \leq u_1(t) \leq \beta(t).$$

Step 5 – The problem (P_n) has at least one solution u_n such that

$$\max(\alpha_1(t), \alpha_2(t), \epsilon_n) \leq u_n(t) \leq u_{n-1}(t).$$

Let us notice that u_{n-1} is an upper solution of (P_n). Since

$$0 = \ddot{u}_{n-1}(t) + f_{n-1}(t, u_{n-1}(t)) \geq \ddot{u}_{n-1}(t) + f_n(t, u_{n-1}(t))$$

and

$$u_{n-1}(0) = u_{n-1}(\pi) = \epsilon_{n-1} \geq \epsilon_n.$$

The claim follows by Theorem 2.6.

Step 6 - Existence of a solution of (3.1). Consider now the pointwise limit

$$\tilde{u}(t) = \lim_{n \to \infty} u_n(t).$$

It is clear that, for any $n \geq 1$,

$$\max(\alpha_1(t), \alpha_2(t)) \leq \tilde{u}(t) \leq u_n(t), \quad \forall t \in]0, \pi[.$$

Using Arzelá-Ascoli Theorem on compact subintervals of $]0, \pi[$, we can prove that $\lim_{n \to \infty} u_n = \tilde{u}$ in $\mathcal{C}^1$. Next, from the closedness of the derivative, we deduce $\tilde{u} \in \mathcal{C}^2(]0, \pi[, \mathbb{R}_0^+)$ and, for all $t \in]0, \pi[$,

$$\ddot{\tilde{u}}(t) + f(t, \tilde{u}(t)) = 0.$$

Since

$$\tilde{u}(0) = \tilde{u}(\pi) = \lim_{n \to \infty} \epsilon_n = 0,$$

it remains only to check the continuity of $\tilde{u}$ at $t = 0$ and $t = \pi$. This can be deduced from the continuity of u_n and the fact that $u_n(0) = u_n(\pi) = \epsilon_n \to 0$. Indeed, for any $\eta > 0$, if n is large enough and t small enough,

$$0 < \tilde{u}(t) \leq u_n(t) \leq \epsilon_n + \eta/2 < \eta. \qquad \blacksquare$$

As a first example, notice that the function

$$f(t, u) = \frac{1}{(\pi - t)t} \frac{1}{u}$$

in (3.3) satisfies the assumptions of the above theorem.

Example 3.1 A model example is the so-called generalized Emden-Fowler equation:

$$\ddot{u} + \frac{f(t)}{u^\sigma} = h(t),$$
$$u(0) = 0, \ u(\pi) = 0,$$

where $\sigma > 0$, $f, h \in \mathcal{C}(]0, \pi[)$, $f > 0$ on $]0, \pi[$ and

$$\int_0^\pi t(\pi - t)(f(t) + |h(t)|) \, dt < +\infty.$$

Notice that in this example h can change sign, f and h can be singular at $t = 0$ and $t = \pi$.

Several extensions can be found in the literature, for example the following problem deals with non homogeneous boundary conditions.

Problem 3.1 Prove the following result.
   Let $f, h \in C(]0, \pi[)$, $f(t) > 0$ on $]0, \pi[$, $\sigma > 0$, $a \geq 0$ and $b \geq 0$. Suppose that

$$\int_0^\pi t(\pi - t)(f(t) + |h(t)|) \, dt < +\infty.$$

Then the problem

$$\ddot{u} + \frac{f(t)}{u^\sigma} = h(t),$$

$$u(0) = a, \; u(\pi) = b,$$

has a solution.
Hint : See J. Janus - J. Myjak [71].

   The next problem concerns condition (iii) in Theorem 3.1.

Problem 3.2 Consider the boundary value problem

$$\ddot{u} + \lambda u(1 + \sin u) + \frac{1}{t(\pi - t)} = 0,$$

$$u(0) = 0, \; u(\pi) = 0,$$

where $\lambda \geq 0$. Notice that we obtain existence of a solution from Theorem 3.1 if $\lambda < 1/2$. Prove that we have existence if $\lambda < 1$.
Hint : See P. Habets and F. Zanolin [64].

   Other results concern regularity at the end points of the interval $[0, \pi]$.

Problem 3.3 Using the set of measurable functions $h : [0, \pi] \to \mathbb{R}^+$ such that

$$\int_0^\pi sh(s) \, ds < +\infty$$

extend Theorem 3.1 to obtain solutions $u \in C^1((0, \pi], \mathbb{R}^+)$.
   Similarly, obtain solutions $u \in C^1([0, \pi), \mathbb{R}^+)$ from the condition

$$\int_0^\pi (\pi - s)h(s) \, ds < +\infty.$$

Hint : See P. Habets and F. Zanolin [63].

3.2 Pairs of positive solutions

In the first section, we have considered assumptions such that the "slope" $\frac{f(t,u)}{u}$ is larger than the first eigenvalue when u goes to zero and smaller than this first eigenvalue when u goes to infinity. This lead to the existence of at least one positive solution. In this section, we consider the case where the "slope" $\frac{f(t,u)}{u}$ is larger than the first eigenvalue near 0 and near infinity. As it is obvious from the example

$$\ddot{u} + 4u = \sin 2t,$$
$$u(0) = 0, \ u(\pi) = 0,$$

these assumptions alone do not guarantee the existence of positive solutions.

To obtain such a result for an elliptic Dirichlet problem, D. G. de Figueiredo and P. L. Lions [40] suppose moreover the existence of a strict upper solution and obtain the existence of two positive solutions. Also, we can prove that the assumption near infinity implies that we have an a priori bound on the upper solutions of (3.1) satisfying the boundary conditions. From this remark, it is possible to apply Theorem 2.16.

Our first result uses an hypothesis of nonresonance type with respect to the first eigenvalue at infinity.

Theorem 3.2 *Let $f :]0, \pi[\times \mathbb{R}_0^+ \to \mathbb{R}$ be a continuous function. Assume there exist α, $\beta \in C([0, \pi])$, respectively a $W^{2,1}$-lower solution and a strict $W^{2,1}$-upper solution, such that $\beta(0) > 0$, $\beta(\pi) > 0$, and for all $t \in]0, \pi[$*

$$0 < \alpha(t) \leq \beta(t).$$

Assume further
(i) for every $R > 0$, there exists $h_R \in \mathcal{A}$ such that, for a.e. $t \in]0, \pi[$ and all $u \in [\alpha(t), R]$,

$$|f(t,u)| \leq h_R(t);$$

(ii) there exist $\rho > 0$ and $b, c \in \mathcal{A}$, $b > 0$ on $[0, \pi]$ such that
• for a.e. $t \in [0, \pi]$ and all $u \geq \rho$,

$$f(t,u) \geq b(t)\,u - c(t);$$

• the first eigenvalue λ_1 of

$$\ddot{u} + \lambda\,bu = 0,$$
$$u(0) = 0, \ u(\pi) = 0,$$

is such that $\lambda_1 < 1$.
Then the problem (3.1) has at least two positive solutions in $C([0, \pi], \mathbb{R}^+) \cap C^2(]0, \pi[, \mathbb{R}_0^+)$.

A similar result can be obtained from a resonance condition.

Theorem 3.3 *Let $f :]0, \pi[\times \mathbb{R}_0^+ \to \mathbb{R}$ be a continuous function. Assume there exist α, $\beta \in \mathcal{C}([0, \pi])$, respectively a $W^{2,1}$-lower solution and a strict $W^{2,1}$-upper solution, such that $\beta(0) > 0$, $\beta(\pi) > 0$, and for all $t \in]0, \pi[$*

$$0 < \alpha(t) \leq \beta(t).$$

Assume further that

(i)   for every $R > 0$, there exists $h_R \in \mathcal{A}$ such that, for a.e. $t \in]0, \pi[$ and all $u \in [\alpha(t), R]$,

$$|f(t, u)| \leq h_R(t);$$

(ii)   there exist $\rho > 0$ and $b \in \mathcal{A}$, $b > 0$ on $[0, \pi]$ such that:
   • for a.e. $t \in [0, \pi]$ and all $u \geq \rho$,

$$f(t, u) \geq b(t) u;$$

   • the first eigenvalue λ_1 of

$$\ddot{u} + \lambda b u = 0,$$
$$u(0) = 0, \ u(\pi) = 0,$$

is such that $\lambda_1 \leq 1$;
   • $f_\infty(t) := \liminf\limits_{u \to +\infty} [f(t, u) - b(t) u] \geq 0$ and $f_\infty \not\equiv 0$.
Then the problem (3.1) has at least two positive solutions in $\mathcal{C}([0, \pi], \mathbb{R}^+) \cap \mathcal{C}^2(]0, \pi[, \mathbb{R}_0^+)$.

The proof of these theorems, which follows from Theorem 2.16, will be omitted but the interested reader can find them in C. De Coster, M. R. Grossinho and P. Habets [35]. It relies on auxiliary results concerning the linear eigenvalue problem

$$\ddot{u} + \lambda b(t) u = 0,$$
$$u(0) = 0, \ u(\pi) = 0, \tag{3.4}$$

in case $b \in \mathcal{A}$.

Remark 3.2 Observe that condition (ii) of Theorem 3.2 generalizes the nonresonance condition

$$\liminf_{u \to +\infty} \frac{f(t, u)}{u} \geq b(t)$$

uniformly in t with $b(t) \geq 1$ and $b(t) > 1$ on a subset of $[0, \pi]$ of positive measure. This condition was introduced by J. Mawhin and J.R. Ward [94] and [95].

Notice also that condition (ii) of Theorem 3.3 is related to the classical Landesman-Lazer condition. In case of resonance, $\lambda_1 = 1$, this condition (ii) implies the Landesman-Lazer condition $\int_0^\pi f_\infty(t)\varphi_1(t)\,dt > 0$, where φ_1 is the eigenfunction of (3.4) corresponding to λ_1.

There are several conditions which give us the existence of the lower and strict upper solutions required in Theorem 3.2 and 3.3. For example, we can use the following propositions which have the advantage to conserve the singular behaviour of f at $t = 0$ and $t = \pi$.

Proposition 3.4 *Let $b \in \mathcal{A}$, $b(t) > 0$ a.e. on $[0, \pi]$ and assume that the first eigenvalue of (3.4) satisfies $\lambda_1 \leq 1$. Suppose that there exists $\epsilon > 0$ such that, for a.e. $t \in]0, \pi[$ and all $u \in]0, \epsilon[$,*

$$f(t, u) \geq b(t)\, u.$$

Then the problem (3.1) has a $W^{2,1}$-lower solution α such that $\alpha(t) \in [0, \epsilon[$ for every $t \in [0, \pi]$.

Proof : Let α be a positive eigenfunction of (3.4) that corresponds to λ_1 and such that $\max_t \alpha(t) < \epsilon$. Then $\alpha(t)$ is a $W^{2,1}$-lower solution of (3.1).   ∎

Remark 3.3 Observe that the condition of Proposition 3.4 are satisfied if

$$\liminf_{u \to 0} \frac{f(t, u)}{u} \geq b(t)$$

uniformly in t with $b(t) \geq 1$ and $b(t) > 1$ on a subset of $[0, \pi]$ of positive measure.

Proposition 3.5 *Let $b \in \mathcal{A}$, $b(t) > 0$ a.e. on $[0, \pi]$ and assume that the first eigenvalue of (3.4) satisfies $\lambda_1 \geq 1$. Suppose that there exist $M > m > 0$ such that, for a.e. $t \in]0, \pi[$ and all $u \in [m, M]$,*

$$f(t, u) \leq b(t)(u - m).$$

Then the problem (3.1) has a strict $W^{2,1}$-upper solution β such that $\beta(t) \in [m, M]$ for every $t \in [0, \pi]$.

Proof : Let β_1 be a positive eigenfunction of (3.4) that corresponds to λ_1 and such that $\max_t \beta_1(t) < M - m$. Then $\beta(t) = \beta_1(t) + m$ is a strict $W^{2,1}$-upper solution of (3.1). ∎

Remark 3.4 In case we can choose $\epsilon < m$ in Propositions 3.4 and 3.5, they provide the existence of a lower and a strict upper solution of problem (3.1) such that $\alpha < \beta$.

Example 3.2 Consider the function $f :]0, \pi[\times \mathbb{R}_0^+ \to \mathbb{R}$ defined by

$$f(t, u) = \frac{3}{t(\pi - t)} \left(\frac{1}{u^2} - \frac{2}{u} + u \right).$$

The main feature of this example is that each of the terms $\frac{1}{u^2}$, $-\frac{2}{u}$, u provides one of the basic assumptions in Theorem 3.3. The function α can be build from the singularity

$\frac{1}{u^2}$ near the origin. The upper solution follows from the term $-\frac{2}{u}$ which dominates on a bounded interval away from the origin. The term u implies the appropriate behaviour at infinity. This is clear if one check the following.
(a) The function

$$a(t) = \alpha_0\, t^{1/3}(\pi - t)^{1/3}$$

is a lower solution if $\alpha_0 > 0$ is small enough.
(b) The constant function

$$\beta(t) = 0,9$$

is a strict upper solution.
(c) Given $R > 0$, if $u \in [\alpha(t), R]$ we have

$$|f(t, u)| = \left| \frac{3}{t(\pi - t)} \left(\frac{1}{u^2} - \frac{2}{u} + u \right) \right|$$

$$\leq \frac{3}{t(\pi - t)} \left(\frac{1}{\alpha_0^2 t^{2/3}(\pi - t)^{2/3}} + \frac{2}{\alpha_0 t^{1/3}(\pi - t)^{1/3}} + \alpha_0\, R \right)$$

$$=: h_R(t)$$

and $h_R \in \mathcal{A}$.
(d) Take $b(t) = \frac{2}{t(\pi-t)}$. Easy computations show that the first eigenvalue of

$$u'' + \lambda b(t)u = 0,$$
$$u(0) = 0,\ u(\pi) = 0$$

is $\lambda_1 = 1$ with eigenfunction $\varphi_1(t) = t(\pi - t)$. Then (ii) is satisfied since

$$\liminf_{u \to +\infty}[f(t, u) - b(t)u] = \liminf_{u \to +\infty} \frac{1}{t(\pi - t)}[3(\frac{1}{u^2} - \frac{2}{u} + u) - 2u] = +\infty.$$

4   An Ambrosetti-Prodi problem

In this chapter, we consider boundary value problems depending on parameters and we study existence and multiplicity results. A model example for the situation considered here is

$$\ddot{u} + 2u_+ = \nu \sin t,$$
$$u(0) = 0,\ u(\pi) = 0, \tag{4.1}$$

where $u_+ = \max(u, 0)$ and $\nu \in \mathbb{R}$. It is easy to see that
   (i) if $\nu < 0$, the problem (4.1) has no solution;
   (ii) if $\nu = 0$, the problem (4.1) has only the trivial solution;
   (iii) if $\nu > 0$, the problem (4.1) has at least two solutions $\pm\nu \sin t$.
This kind of result is classically called an Ambrosetti-Prodi problem.

In 1972, A. Ambrosetti and G. Prodi [12] have considered the problem of characterizing the set of functions h such that the Dirichlet problem

$$\Delta u + f(u) = h(x), \text{ in } \Omega,$$
$$u = 0, \text{ on } \partial\Omega,$$

has a solution. They have investigate the case of a convex function f which satisfies

$$0 < \lim_{u \to -\infty} f'(u) < \lambda_1 < \lim_{u \to +\infty} f'(u) < \lambda_2,$$

where λ_1, λ_2 are the two first eigenvalues of the Laplacian operator together with Dirichlet boundary conditions. From such an assumption, they obtained a manifold M which separates the space of forcings h in two regions O_0 and O_2 such that the above problem has zero, exactly one or exactly two solutions according to $h \in O_0$, $h \in M$ or $h \in O_2$. Later, R. Chiappinelli, J. Mawhin and R. Nugari [27] have investigated the problem

$$\ddot{u} + u + f(t, u) = \nu\varphi(t),$$
$$u(0) = 0, \ u(\pi) = 0,$$
(4.2)

where $\varphi(t) = \sqrt{\frac{2}{\pi}} \sin t$. Using a coercitivity assumption

$$\lim_{|u| \to +\infty} f(t, u) = +\infty,$$

uniformly in t, they proved there exist ν_0 and $\nu_1 > \nu_0$ such that
(i) if $\nu < \nu_0$, the problem (4.2) has no solution;
(ii) if $\nu_0 \leq \nu \leq \nu_1$, the problem (4.2) has at least one solution;
(iii) if $\nu_1 < \nu$, the problem (4.2) has at least two solutions.
Under the additional condition that $f(t, u) + Ku$ is nondecreasing for u in some neighbourhood of the origin, they proved that $\nu_0 = \nu_1$. J. L. Kazdan and F. W. Warner [74] remarked that $\nu_0 = \nu_1$ if $\varphi(t) = 1$ and more generally whenever φ is one sign. The problem is to recognize if an upper solution is strict or not in order to apply Theorem 2.15.

In order better to understand the relation between these two results, we consider in this chapter the two parameters problem

$$\ddot{u} + u + f(t, u) = \mu + \nu\varphi(t),$$
$$u(0) = 0, \ u(\pi) = 0.$$
(4.3)

Detailed proofs of the theorems, together with additional results and examples, can be found in [36].

4.1   Existence results

A first result separates the space of parameters (μ, ν) in two regions O_0 and O_1 such that the problem (4.3) has no solution in O_0 and at least one in O_1.

Theorem 4.1 *Let f be a L^p-Carathéodory function such that*
(H-1)for any $t_0 \in [0, \pi]$ and any bounded set $E \subset \mathbb{R}$, there exists an interval I_0 with $t_0 \in I_0$ and

$$\forall \epsilon > 0, \exists \delta > 0, \ for \ a.e. \ t \in I_0, \forall y \in E, \forall x \in [y - \delta \sin t, y]$$
$$f(t, x) - f(t, y) \le \epsilon;$$

(H-2) $\lim_{u \to -\infty} f(t, u) = +\infty$, *uniformly in* t
and
(H-3)there exists $k \in L^1(0, \pi)$ such that for a.e. $t \in [0, \pi]$ and all $u \in \mathbb{R}$, $f(t, u) \ge k(t)$.
Then, there exists a nonincreasing, Lipschitz function $\mu_0 : \mathbb{R} \to \mathbb{R}$ such that
(i) if $\mu < \mu_0(\nu)$, the problem (4.3) has no solution;
(ii) if $\mu_0(\nu) < \mu$, the problem (4.3) has at least one solution in $W^{2,1}(0, \pi)$.

Remark 4.1 Assumption (H-1) is an upper semicontinuity of f from the left in x with some uniformity in t.

Typical examples of L^p-Carathéodory functions satisfying (H-1) are

$$f(t, u) = g(u) + h(t),$$

with g continuous and $h \in L^p(0, \pi)$.

The condition (H-1) implies that a solution β of (4.3) with $(\mu, \nu) = (\mu_0, \nu_0)$ is a strict upper solution of (4.3) for any (μ, ν) such that

$$\mu > \mu_0 \quad and \quad \mu + \sqrt{\frac{2}{\pi}} \nu > \mu_0 + \sqrt{\frac{2}{\pi}} \nu_0.$$

Sketch of the proof. First part – Let us fix $\nu \in \mathbb{R}$ and prove the existence of $\mu_0(\nu)$.
Step 1 – Claim : The problem (4.3) has no solution for μ negative enough. Multiplying the equation by φ and integrating, one obtains a lower bound on μ.
Step 2 – Claim : There exists μ large enough so that the problem (4.3) has a strict upper solution β. Let $m \in L^1(0, \pi)$ be such that for a.e. $t \in [0, \pi]$ and all $u \in [-1, 1]$, $|f(t, u)| \le m(t)$. For $h \in L^1(0, \pi)$ denote $\overline{h} = \int_0^\pi h(t)\varphi(t)\, dt$. Choose $p \in C([0, \pi])$, a L^1–approximation of m, close enough so that the problem

$$\ddot{u} + u + (m - p) - (\overline{m} - \overline{p})\varphi(t) = 0,$$
$$u(0) = 0, \quad u(\pi) = 0$$

has a solution $\beta \in [-1, 1]$. For μ large enough, β is a strict upper solution.
Step 3 – Claim : The problem (4.3) (with μ as in Step 2) has a strict lower solution $\alpha \prec \beta$. From the assumptions, we have $k \in L^1(0, \pi)$ and we can find $r > 0$ such that

$$f(t, u) \ge k(t), \qquad\qquad if \ u \in \mathbb{R},$$
$$\ge \mu + \frac{\nu}{2}\sqrt{\frac{\pi}{2}} + 1, \quad if \ u < -r.$$

Let $F_\delta := [0, \delta] \cup [\pi - \delta, \pi]$ and define

$$g(t) = k(t) - \mu, \qquad \text{if } t \in F_\delta,$$
$$= \tfrac{\nu}{2}\sqrt{\tfrac{\pi}{2}} + 1, \qquad \text{if } t \notin F_\delta.$$

We can choose δ small enough so that

$$\bar{g} = \int_0^\pi g(t)\varphi(t)\, dt > \nu.$$

Define now w to be the solution of

$$\ddot{w} + w + g(t) - \bar{g}\varphi(t) = 0,$$
$$w(0) = 0, \quad w(\pi) = 0.$$

For a negative enough, the function

$$\alpha(t) = a\varphi(t) + w(t)$$

verifies the claim.

Step 4 – We define

$$\mu_0(\nu) = \inf\{\mu \mid (4.3) \text{ has a solution for } (\mu, \nu)\}.$$

From the previous steps and Theorem 2.4, $\mu_0(\nu)$ exists.

Step 5 – Claim : For $\mu > \mu_0(\nu)$ the problem (4.3) has at least one solution. From the definition of $\mu_0(\nu)$, given $\mu > \mu_0(\nu)$, there exists $\mu_1 \in [\mu_0(\nu), \mu[$ and u_1, solution of (4.3) for (μ_1, ν). This function u_1 is a strict upper solution of (4.3) for (μ, ν). The existence of a strict lower solution follows as in step 3. The assertion follows then from Theorem 2.4.

Second part – The function μ_0 is nonincreasing and Lipschitz. Let $\nu_1 < \nu_2$. Using the arguments of step 5, we prove there exists a strict upper solution β if (μ_1, ν_1) is such that $\mu_1 > \mu_0(\nu_2) + \sqrt{\tfrac{2}{\pi}}(\nu_2 - \nu_1)$. From the argument of step 3, we find a strict lower solution $\alpha \prec \beta$ and from Theorem 2.4, it is clear that $\mu_0(\nu_1) \le \mu_0(\nu_2) + \sqrt{\tfrac{2}{\pi}}(\nu_2 - \nu_1)$. Using a similar idea, we prove that $\mu_0(\nu_1) \ge \mu_0(\nu_2)$. ∎

4.2 Multiplicity results

Consider the problem (4.3) with

$$f(t, u) := u^-$$

It is easy to see that $\mu_0(0) = 0$. First, one proves there is no solution if $\nu = 0$ and $\mu < 0$. Next, considering a solution for $\nu = 0$ and $\mu > 0$, the distance between two consecutive zeros of a positive hump of such a solution is larger than π. Hence, this solution has to be negative and is unique. This means that for $\nu = 0$, Theorem 4.1 gives an exact count of the number of solutions. To obtain a multiplicity result, we will have to reinforce (H-2) and (H-3), assuming a better control on the nonlinearity for large values of u.

Theorem 4.2 *Let f be a L^p-Carathéodory function which satisfies (H-1) and*
(H-4) $\lim\limits_{|u| \to +\infty} f(t, u) = +\infty$, *uniformly in t.*
Then there exists a nonincreasing, Lipschitz function $\mu_0 : \mathbb{R} \to \mathbb{R}$ such that
(i) if $\mu < \mu_0(\nu)$, the problem (4.3) has no solution;
(ii) if $\mu = \mu_0(\nu)$, the problem (4.3) has at least one solution in $W^{2,p}(0, \pi)$;
(iii) if $\mu_0(\nu) < \mu$, the problem (4.3) has at least two solutions in $W^{2,p}(0, \pi)$.

Sketch of the proof. Let $\mu_0(\nu)$ be defined from Theorem 4.1 and consider ν_1 and $\mu_1 > \mu_0(\nu_1)$.

Step 1 – Claim : There is $r > 0$ such that for all $\mu < \mu_1$ and all solutions u of (4.3) with (μ, ν_1), we have $\|u\|_\infty < r$. Let u be such a solution. It is proved in [88] that for some $K > 0$ independent of u, one has

$$\|\tilde{u}\|_\infty \le K \int_0^\pi |\ddot{\tilde{u}}(t) + \tilde{u}(t)| \varphi(t) \, dt,$$

where $u = \tilde{u} + \bar{u}\varphi$ and $\bar{u} = \int_0^\pi u\varphi \, dt$, Also, there exists some function $m \in L^1(0, \pi)$ such that

$$|\ddot{u}(t) + u(t)| \le f(t, u(t)) + m(t) + |\mu| + |\nu_1|\varphi(t).$$

Hence, we obtain some $\tilde{r} \in \mathbb{R}$ such that $\|\tilde{u}\|_\infty \le \tilde{r}$.

At last, we assume there exists a sequence $(u_n)_n$ such that $\|u_n\|_\infty \to +\infty$. If we write $u_n = \tilde{u}_n + \bar{u}_n\varphi$ as above, we must have $|\bar{u}_n| \to +\infty$ and, using Fatou's theorem, we come to a contradiction. This proves there exists $r \in \mathbb{R}$ such that $\|u\|_\infty \le r$ and the claim follows.

Step 2 – We can find as in Theorem 4.1 strict lower and upper solutions $\alpha \prec \beta$. Hence, we can apply Theorem 2.15 which proves that for (μ_1, ν_1) the problem (4.3) has at least two solutions.

Step 3 – Claim : If $\mu = \mu_0(\nu)$, the problem (4.3) has at least one solution. Consider a sequence $(\mu_n)_n$ such that $\mu_n > \mu_0(\nu)$ and $\mu_n \to \mu_0(\nu)$, together with a solution u_n of (4.3) for (μ_n, ν). From Arzela-Ascoli Theorem, it is easy to find subsequences such that u_n converges to a function u solution of (4.3) for (μ, ν). ∎

If we recall that μ_0 is nonincreasing, we can summarize Theorem 4.2 in Figure 1. It is then obvious that, for μ constant, for example $\mu = 0$, Theorem 4.2 reduces to the following result which extends [27] to L^p-Carathéodory functions.

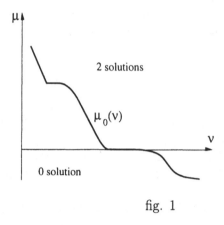

fig. 1

Corollary 4.3 *Let f be a L^p-Carathéodory function such that (H-1) and (H-4) are satisfied.*

Then there exist ν_0 and $\nu_1 \geq \nu_0$ such that
(i) if $\nu < \nu_0$, the problem (4.2) i.e.

$$\ddot{u} + u + f(t, u) = \nu\varphi(t),$$
$$u(0) = 0,\ u(\pi) = 0,$$

has no solution;
(ii) if $\nu_0 \leq \nu \leq \nu_1$, the problem (4.2) has at least one solution in $W^{2,p}(0, \pi)$;
(iii) if $\nu_1 < \nu$, the problem (4.2) has at least two solutions in $W^{2,p}(0, \pi)$.

At last, we can complement this corollary noticing that generically $\nu_0 = \nu_1$. Indeed, this holds for arbitrary small perturbations of f.

4.3   Decreasing of μ_0

In this section, we give some assumptions on f which imply that the function μ_0 given by Theorem 4.1 is decreasing.

Proposition 4.4 *Let f be a L^p-Carathéodory function which satisfies (H-1) and (H-4). Assume one of the functions $g(t, u) = f(t, u)$ or $g(t, u) = -f(t, u)$ is such that (H-5) given $R > 0$, we have*

$$\forall \epsilon > 0, \exists \delta > 0, \text{for a.e. } t \in [0, \pi], (\forall x : |x| \leq R\varphi(t)), (\forall y : |y| \leq R\varphi(t)),$$
$$0 \leq x - y \leq \delta\varphi(t) \Rightarrow g(t, y) - g(t, x) \leq \epsilon\varphi(t).$$

Then, the function $\mu_0(\nu)$ is decreasing.

Proof : Let $\nu_0 \in \mathbb{R}$.

Step 1 – *Claim* : For all $\nu > \nu_0$, there exist $\mu_1 < \mu_0(\nu_0)$ and a strict upper solution β of (4.3) for $\mu \in]\mu_1, \mu_0(\nu_0)]$ and ν.

Assume $g(t, u) = f(t, u)$ in (H-5). The proof is similar in case $g(t, u) = -f(t, u)$. From Theorem 4.2, there is a solution u_0 of

$$\ddot{u}_0 + u_0 + f(t, u_0) = \mu_0(\nu_0) + \nu_0\varphi(t),$$
$$u_0(0) = 0, \ u_0(\pi) = 0.$$

Also, there exists $\widetilde{K} > 0$ such that for every $m > 0$, the solution w of

$$\ddot{w} + w = -m(1 - 2\sqrt{\tfrac{2}{\pi}}\varphi(t)),$$
$$w(0) = 0, \ w(\pi) = 0,$$

with $\overline{w} = 0$, is such that, for all $t \in [0, \pi]$,

$$|w(t)| \leq m\widetilde{K}\varphi(t).$$

Define

$$\beta(t) := u_0(t) + w(t) - m\widetilde{K}\varphi(t),$$

where $m \in]0, 1]$ will be chosen later. Observe that

$$0 \leq u_0(t) - \beta(t) \leq 2m\widetilde{K}\varphi(t).$$

Let $R > 0$ be such that, for all $m \in]0, 1]$,

$$|u_0(t)| < R\varphi(t), \quad |\beta(t)| < R\varphi(t).$$

Define $\epsilon > 0$ such that

$$\nu_0 + \epsilon < \nu$$

and pick $\delta > 0$ from assumption (H-5). Next, we choose $m \in]0, 1]$ small enough so that

$$2m\widetilde{K} \leq \delta, \quad \nu_0 + 2\sqrt{\frac{2}{\pi}}m + \epsilon \leq \nu.$$

and set $\mu_1 := \mu_0(\nu_0) - m$. For these values, we can prove that β is a strict upper solution of (4.3) for $\mu \in]\mu_1, \mu_0(\nu_0)]$ and ν.

Step 2 – It is now easy to conclude as in Step 5 of Theorem 4.1. ∎

Assumption (H-5) can be thought of as some one-sided uniform continuity assumption when $t \neq 0, \pi$, together with a Lipschitz condition near zero. More precisely, we have the following result whose proof can be found in [36].

Proposition 4.5 *Let $f(t, u)$ be a continuous function which verifies*
(H-5') there exist $L > 0$ and $r > 0$ such that

$$f(t, u) + L u \quad or \quad - f(t, u) + L u$$

is nondecreasing in u for a.e. $t \in [0, \pi]$ and $u \in [-r, r]$.
Then assumption (H-5) holds.

Theorem 4.4 implies that the set of points ν so that the problem (4.2) has exactly one solution reduces at most to one point. This is the aim of the following Theorem which improves a result of P. Habets and P. Omari [61] and the corresponding one in R. Chiappinelli, J. Mawhin and R. Nugari [27].

Theorem 4.6 *Let f be a L^p-Carathéodory function such that (H-1), (H-4) and (H-5) are satisfied.*
Then there exists $\nu_0 \in \mathbb{R}$ such that
(i) for $\nu < \nu_0$, the problem (4.2) has no solution;
(ii) for $\nu = \nu_0$, the problem (4.2) has at least one solution in $W^{2,p}(0, \pi)$;
(iii) for $\nu > \nu_0$, the problem (4.2) has at least two solutions in $W^{2,p}(0, \pi)$.

Let us notice at last that similar results are obtained in [36] in case the Lipschitz condition is replaced by a Hölder one.

Problem 4.1 Under which conditions Corollary 4.3 gives an exact count of the solutions?
Hint : See H. Berestycki [16].

Problem 4.2 Consider the problem

$$\ddot{u} + u + f(u) = s\varphi(t),$$
$$u(0) = 0, \ u(\pi) = 0,$$

$$(4.4)$$

where f is a continuous function. Prove that if

$$\lim_{|u| \to +\infty} f(u) = +\infty$$

and, moreover

$$\lambda_2 - 1 = 3 < \lim_{u \to +\infty} \frac{f(u)}{u} =: \mu < 8 = \lambda_3 - 1, \qquad (4.5)$$

then, for s large enough, the problem (4.4) has at least three solutions.
Hint : See A.C. Lazer - P.J. McKenna [79].

Combining the method of lower and upper solutions with a shooting method, we can extend this result.

Problem 4.3 Prove the following result.

Let $f \in C^1(\mathbb{R})$ and $h \in C([0, \pi])$. Assume that, for some $n \geq 1$,

$$\liminf_{u \to -\infty} \frac{f(u)}{u} < 1 \text{ and } n^2 < \lim_{u \to +\infty} f'(u) < (n+1)^2.$$

Then there exists s_1 such that, for $s > s_1$ the problem

$$\ddot{u} + f(u) = h(t) + s \sin t,$$
$$u(0) = 0, \; u(\pi) = 0,$$

has at least $2n$ solutions.

Hint : See A.C. Lazer - P.J. McKenna [80] and [81].

A similar type of result appears for a boundary value problem with a parameter s in the boundary conditions rather than in the forcing. This is described in the following problem.

Problem 4.4 Consider the problem

$$\ddot{u} + f(u) = h(t),$$
$$u(0) = s, \; u(\pi) = sa, \tag{4.6}$$

with $a > 0$ and prove the following results.

Let $h \in C([0, \pi])$ and assume $f \in C(\mathbb{R})$ satisfies

$$\lim_{u \to -\infty} \frac{f(u)}{u} = \alpha, \; \lim_{u \to +\infty} \frac{f(u)}{u} = \beta,$$

with $\alpha < 1 < \beta$. Then there exist $s_- < 0$, $s_+ > 0$ such that, if $s \leq s_-$ the problem (4.6) has at least two solutions and if $s \geq s_+$ the problem (4.6) has no solution.

Moreover, if $f(u) = -\alpha u^- + \beta u^+$ then (4.6) has only the trivial solution if $s = 0$, no solution when $s > 0$ and a unique negative solution if $s < 0$.

Hint : See G. Harris [65].

5 Non-ordered upper and lower solutions

Let us consider the periodic problem

$$\ddot{u} + f(t, u) = 0,$$
$$u(0) = u(2\pi), \; \dot{u}(0) = \dot{u}(2\pi). \tag{5.1}$$

We want to discuss the situation where α and β do not satisfy the "usual" ordering condition

$$\alpha(t) \leq \beta(t) \text{ for all } t \in [0, 2\pi].$$

It was pointed out in [9] (see also Remark 2.2) that, in general, the mere existence of a lower solution and an upper solution for which

$$\alpha(t) \geq \beta(t) \quad \text{for all } t \in [0, 2\pi] \tag{5.2}$$

holds is not sufficient to guarantee the solvability of (5.1).

Lower and upper solutions α, β satisfying the opposite ordering condition (5.2) arise naturally in situations where the corresponding problem has a solution. As a very simple example, one can consider the linear system

$$\ddot{u} + \tfrac{2}{3}u = \sin t,$$
$$u(0) = u(2\pi), \ \dot{u}(0) = \dot{u}(2\pi).$$

The functions $\alpha = \tfrac{3}{2}$ and $\beta = -\tfrac{3}{2} < \alpha$ are lower and upper solutions such that $\alpha \geq \beta$. Notice, however, that the unique solution $u(t) = -3 \sin t$ does not lie between the upper and the lower solution.

More generally, one can consider the problems

$$\ddot{u} + \lambda u = \sin t,$$
$$u(0) = u(2\pi), \ \dot{u}(0) = \dot{u}(2\pi), \tag{5.3}$$

where λ lies between the two first eigenvalues of the periodic problem

$$\ddot{u} + \lambda u = 0,$$
$$u(0) = u(2\pi), \ \dot{u}(0) = \dot{u}(2\pi),$$

i.e. $0 < \lambda < 1$. It is easy to see that, in such a case, there exist constant upper and lower solutions of (5.3) in the reversed order. Further, the solution of (5.3) lies between all of them, $\beta(t) \leq u(t) \leq \alpha(t)$, if and only if $0 < \lambda \leq \tfrac{1}{2}$. We shall see that if we reinforce this condition, i.e. $\lambda \leq \tfrac{1}{4}$, solutions can be obtained using monotone iterations.

A last remark is due to J. Kazdan and F. Warner [74]. Assume there exists upper and lower solutions of (5.1) in the reversed order (5.2). This imposes some conditions on the nonlinearity. Indeed, if we assume further the function $f(t, .)$ is nonincreasing, we can compute

$$0 \geq \dot{\alpha}(2\pi) - \dot{\beta}(2\pi) - \dot{\alpha}(0) + \dot{\beta}(0) = \int_0^{2\pi} (\ddot{\alpha}(t) - \ddot{\beta}(t)) \, dt$$
$$\geq \int_0^{2\pi} (f(t, \beta(t)) - f(t, \alpha(t))) \, dt \geq 0$$

and the equality holds. This implies $\alpha = \beta$ and this function is already a solution of (5.1). For more general problems, the key assumption concerns the nonincreasingness of the function $f(t, x) - \lambda_1 x$, i.e.

$$\frac{f(t, x) - f(t, y)}{x - y} \leq \lambda_1,$$

where λ_1 is the first eigenvalue of the corresponding eigenvalue problem (here $\lambda_1 = 0$). Hence, such upper and lower solutions seem inappropriate to deal with in case f lies in some sense at the left of the first eigenvalue λ_1. Remark also that a similar statement holds for well-ordered upper and lower solutions. If $\alpha \leq \beta$, the function f cannot lie at the right of the first eigenvalue.

The last remark suggests that existence of upper and lower solutions satisfying (5.2) forces the nonlinearity f to be on the right of the first eigenvalue. In order to achieve the solvability, one should prevent the interference of f with the higher part of the spectrum. This can be expressed in various ways.

In the first paragraph, we prove that the existence of a lower solution α and of an upper solution β (satisfying no ordering condition) implies the solvability of (5.1), provided that $f(t, u)$ is bounded. In the second paragraph, we study systems with asymmetric nonlinearities. These are non-selfadjoint problems. Here, the upper and lower solutions are supposed ordered, i.e. they satisfy (5.2). The benefit of this assumption is that it gives some localization of the solution in the sense that there exists $t_0 \in [0, \pi]$ such that $\beta(t_0) < u(t_0) < \alpha(t_0)$. Such a property can be used to obtain multiplicity results. The last paragraph extends some of the ideas on monotone iterations.

5.1   Upper and lower solutions without ordering

Consider the periodic problem (5.1).

Theorem 5.1 *Assume the function* $f : [0, 2\pi] \times \mathbb{R} \to \mathbb{R}$ *is continuous and bounded. Assume further there exist a lower solution* $\alpha(t)$ *and a upper solution* $\beta(t)$.
Then, problem (5.1) has at least one solution.

Proof : Consider the space

$$X = \{u \in C([0, 2\pi]) \mid \int_0^{2\pi} u(t)\, dt = 0\}$$

with the norm $\| \cdot \|_\infty$. Define the compact solution operator

$$K : X \to X$$

such that for any $f \in X$, $u = Kf$ is the unique solution in X of

$$\ddot{u} + f(t) = 0,$$
$$u(0) = u(2\pi),\ \dot{u}(0) = \dot{u}(2\pi)$$

and the continuous operators

$$\overline{N} : C([0, 2\pi]) \to \mathbb{R},$$
$$\widetilde{N} : C([0, 2\pi]) \to X,$$

which are defined by

$$Nu = \frac{1}{2\pi} \int_0^{2\pi} f(t, u(t)) \, dt,$$

and

$$(\widetilde{N}u)(t) = f(t, u(t)) - \overline{N}u.$$

Given $u \in \mathcal{C}([0, 2\pi])$, we write

$$\overline{u} = \frac{1}{2\pi} \int_0^{2\pi} u(t) \, dt \quad \text{and} \quad \widetilde{u} = u - \overline{u}.$$

It is then easy to see that $u \in \mathcal{C}([0, 2\pi])$ is a solution of (5.1) if and only if $\overline{u} \in \mathbb{R}$ and $\widetilde{u} \in X$ are solutions of

$$\widetilde{u} = K\widetilde{N}(\overline{u} + \widetilde{u}),$$
$$\overline{N}(\overline{u} + \widetilde{u}) = 0.$$

Step 1 - Claim : The set

$$\Sigma = \{(\overline{u}, \widetilde{u}) \in \mathbb{R} \times X \mid \widetilde{u} = K\widetilde{N}(\overline{u} + \widetilde{u})\}$$

is such that for any $a > 0$, there exists a connected subset Σ_a of Σ with $\mathrm{proj}_{\mathbb{R}}\Sigma_a :=$ $\{\overline{u} \mid \exists \widetilde{u} \in X, (\overline{u}, \widetilde{u}) \in \Sigma_a\} = [-a, a]$.

Observe that for any $\overline{u}$ the operator $T\widetilde{u} = K\widetilde{N}(\overline{u} + \widetilde{u})$ is bounded, completely continuous. Hence by Schauder's Theorem, there exists $\widetilde{u} \in X$ such that $\widetilde{u} = T\widetilde{u}$. This implies $\mathrm{proj}_{\mathbb{R}}\Sigma = \mathbb{R}$. Notice also that the boundedness of T implies there exists $R > 0$ such that $\Sigma \subset \mathbb{R} \times B_R$.

Define now $K := \{(\overline{u}, \widetilde{u}) \in \Sigma \mid -a \leq \overline{u} \leq a\}$ and $K_r := \{(\overline{u}, \widetilde{u}) \in \Sigma \mid \overline{u} = r\}$. The set K is compact and $K_{\pm a}$ are nonempty disjoint, closed subsets of K.

Suppose there does not exist a connected subset Σ_a of K joining K_{-a} and K_a. Then there are disjoint compact subsets $C_a \supset K_a$ and $C_{-a} \supset K_{-a}$, with $K = C_{-a} \cup C_a$. Hence we can find an open set $\Omega \subset\,]-\infty, a[\times B_R$ such that

$$C_{-a} \subset \Omega \quad \text{and} \quad C_a \cap \overline{\Omega} = \emptyset.$$

Let $\Omega_r := \{\widetilde{u} \in X \mid r + \widetilde{u} \in \Omega\}$. It follows then from the properties of the degree (see e.g. [22]) that

$$1 = \deg(I - K\widetilde{N}(\cdot - a), B_R) = \deg(I - K\widetilde{N}(\cdot - a), \Omega_{-a})$$
$$= \deg(I - K\widetilde{N}(\cdot + a), \Omega_a) = 0,$$

which is a contradiction.

Step 2 - If there exists $(\overline{u}, \widetilde{u}) \in \Sigma$ such that $\overline{N}(\overline{u} + \widetilde{u}) = 0$, it is clear that $u = \overline{u} + \widetilde{u}$ is a solution of (5.1).

Let us prove that if such a $(\overline{u}, \widetilde{u})$ does not exist (i.e. for any $(\overline{u}, \widetilde{u}) \in \Sigma$, $\overline{N}(\overline{u} + \widetilde{u}) \neq 0$), we can find well-ordered upper and lower solutions. For any $a > 0$, define Σ_a from the claim in Step 1 and since $\overline{N}$ is continuous, the set $\{\overline{N}(\overline{u} + \widetilde{u}) \mid (\overline{u}, \widetilde{u}) \in \Sigma_a\}$ is an interval that does not contain zero. Let us assume $\overline{N}(\overline{u} + \tilde{u}) > 0$. In that case, any function $u = \overline{u} + \tilde{u}$ such that $(\overline{u}, \tilde{u}) \in \Sigma_a$ is a lower solution since

$$\ddot{u} + f(t, u) = \ddot{\tilde{u}} + f(t, \overline{u} + \tilde{u}) = \overline{N}(\overline{u} + \tilde{u}) > 0,$$
$$u(0) = u(2\pi), \ \dot{u}(0) = \dot{u}(2\pi).$$

If we pick a large enough so that

$$R - a < \beta(t),$$

where R is such that $\Sigma \subset \mathbb{R} \times B_R$, and $\tilde{u}$ so that $(-a, \tilde{u}) \in \Sigma_a$, we have a lower solution

$$\alpha_0(t) = \tilde{u}(t) - a < \beta(t).$$

Similarly, if $\overline{N}(\overline{u} + \tilde{u}) < 0$, we build an upper solution $\beta_0 > \alpha$. The existence follows now from Theorem 2.2. ∎

As an application, consider the following result.

Theorem 5.2 *Let $f : [0, 2\pi] \times \mathbb{R} \to \mathbb{R}$ be a continuous, bounded function and $h \in C([0, 2\pi])$. Assume that for some $R > 0$ and all $(t, x) \in \{(t, x) \in [0, 2\pi] \times \mathbb{R} \mid |x| \geq R\}$,*

$$f(t, x) \operatorname{sgn} x \geq 0,$$

and

$$\int_0^{2\pi} h(t) \, dt = 0.$$

Then the problem

$$\ddot{u} + f(t, u) = h(t),$$
$$u(0) = u(2\pi), \ \dot{u}(0) = \dot{u}(2\pi)$$

has at least one solution.

Proof : Let us show how to construct a positive lower solution. Take a solution w of

$$\ddot{u} = h(t),$$
$$u(0) = u(2\pi), \ \dot{u}(0) = \dot{u}(2\pi)$$

and choose s large enough so that $s + w(t) \geq R$ in $[0, 2\pi]$. It is then easy to see that $\alpha(t) = s + w(t)$ is the required lower solution.

In the same way, we construct an upper solution $\beta(t) < -R$. The result follows then from Theorem 5.1. ∎

Example 5.1 Consider the following problem

$$\ddot{u} + u \exp(-u^2) = h(t),$$
$$u(0) = u(2\pi), \ \dot{u}(0) = \dot{u}(2\pi)$$

where $h \in C([0, 2\pi])$ is such that $\int_0^{2\pi} h(t) \, dt = 0$. Notice that the existence of a solution follows from the above theorem but cannot be deduced from Landesman-Lazer conditions such as in Proposition 5.8 below.

In the next example, we use upper and lower solutions u_1 and u_2 which are not ordered.

Example 5.2 Consider the problem

$$\ddot{u} + g(u) = \overline{h} + \tilde{h}(t),$$
$$u(0) = u(2\pi), \ \dot{u}(0) = \dot{u}(2\pi), \tag{5.4}$$

where $g : \mathbb{R} \to \mathbb{R}$ is a bounded continuous function and let us prove that, given $\tilde{h} \in L^1(0, 2\pi)$ with $\int_0^{2\pi} \tilde{h}(t) \, dt = 0$, the values of $\overline{h} \in \mathbb{R}$ for which the problem (5.4) has a solution is an interval. To see that, let $\overline{h}_1 < \overline{h}_2$ be such that the problems

$$\ddot{u} + g(u) = \overline{h}_i + \tilde{h}(t),$$
$$u(0) = u(2\pi), \ \dot{u}(0) = \dot{u}(2\pi),$$

have a solution u_i. Let $\overline{h} \in [\overline{h}_1, \overline{h}_2]$ then u_1 and u_2 are upper and lower solutions of (5.4). The result follows from Theorem 5.1.

5.2 Upper and lower solutions in the reversed order

In this section, we consider the problem (5.1) for asymmetric nonlinearities, i.e. we assume there exist functions $a, b, c, d \in L^1(0, 2\pi)$ such that

$$a(t) \leq \liminf_{u \to +\infty} \frac{f(t, u)}{u} \leq \limsup_{u \to +\infty} \frac{f(t, u)}{u} \leq b(t),$$
$$c(t) \leq \liminf_{u \to -\infty} \frac{f(t, u)}{u} \leq \limsup_{u \to -\infty} \frac{f(t, u)}{u} \leq d(t), \tag{5.5}$$

uniformly in t. Further, we will assume the nonlinearity f to be a L^1-Carathéodory function.

An important assumption concerns the control of the linear growth of f. This will be done from the following definition.

Definition 5.1 *Let $a, b, c, d \in L^1(0, 2\pi)$ be such that $a(t) \leq b(t), c(t) \leq d(t)$ a.e. in $[0, 2\pi]$. The set*

$$[a, b] \times [c, d] = \{(p, q) \in L^1(0, 2\pi) \times L^1(0, 2\pi) \mid$$
$$a(t) \leq p(t) \leq b(t), c(t) \leq q(t) \leq d(t), \ a.e. \ in \ [0, 2\pi]\}$$

is said to be admissible *if the following conditions hold.*
(A) For any $(p, q) \in [a, b] \times [c, d]$, any nontrivial solution u of

$$\ddot{u} + p(t)u^+ - q(t)u^- = 0,$$
$$u(0) = u(2\pi), \ \dot{u}(0) = \dot{u}(2\pi), \tag{5.6}$$

where $u^+(t) = \max(u(t), 0)$ and $u^-(t) = \max(-u(t), 0)$, is one-signed on $[0, 2\pi]$, i.e.

$$u(t) > 0 \ for \ all \ t \in [0, 2\pi] \ or \ u(t) < 0 \ for \ all \ t \in [0, 2\pi].$$

(B) *There exist two continuous functions* $\mu \in [0,1] \to p^\mu, q^\mu \in L^1(0, 2\pi)$ *such that*
  - $(p^1, q^1) \in [a, b] \times [c, d]$ *and* $p^1(t), q^1(t) \geq 0$ *a.e. in* $[0, 2\pi]$;
  - $p^0 = q^0$;
  - *for any* $\mu \in [0, 1]$, *the problem*

$$\ddot{u} + p^\mu(t)u^+ - q^\mu(t)u^- = 0,$$
$$u(0) = u(2\pi), \; \dot{u}(0) = \dot{u}(2\pi),$$

has only the trivial solution.

The set F of points $(\mu, \nu) \in \mathbb{R}^2$ such that the problem

$$\ddot{u} + \mu u^+ - \nu u^- = 0,$$
$$u(0) = u(2\pi), \; \dot{u}(0) = \dot{u}(2\pi), \tag{5.7}$$

has a nontrivial solution is known as the Fučík spectrum of (5.7). From an explicit computation of the solutions of (5.7), it is easy to see that

$$F = \bigcup_{n=1}^{\infty} F_n,$$

where

$$F_1 = \{(\mu, 0) \mid \mu \in \mathbb{R}\} \cup \{(0, \nu) \mid \nu \in \mathbb{R}\}$$

and

$$F_n = \{(\mu, \nu) \mid \frac{1}{\sqrt{\mu}} + \frac{1}{\sqrt{\nu}} = \frac{2}{n-1}\}, \quad n = 2, 3, \ldots$$

The following lemma can be used to describe sets that are admissible.

Lemma 5.3 *Let* $(\mu, \nu) \in F_2$ *and* $p, q \in L^1(0, 2\pi)$. *Assume there exists some set* $I \subset [0, 2\pi]$ *of positive measure such that*

$$p(t) \leq \mu, \; q(t) \leq \nu, \quad \text{for a.e. } t \in [0, 2\pi],$$
$$p(t) < \mu, \; q(t) < \nu, \quad \text{for a.e. } t \in I. \tag{5.8}$$

Then problem (5.6) has only one-signed solutions.

Proof : Assume there exists a solution u which is not one-signed. Extend u by periodicity and let t_0 and t_1 be consecutive zeros such that u is positive on $]t_0, t_1[$. Define $v(t) = \sin(\sqrt{\mu}(t - t_0))$ and compute

$$(u\dot{v} - v\dot{u})\Big|_{t_0}^{t_1} = \int_{t_0}^{t_1} (p(t) - \mu)u(t)v(t)\, dt \leq 0.$$

If $t_1 - t_0 < \frac{\pi}{\sqrt{\mu}}$, we come to a contradiction

$$0 < -v(t_1)\dot{u}(t_1) \leq 0.$$

Hence, $t_1 - t_0 \geq \frac{\pi}{\sqrt{\mu}}$ and we only have equality in case $p(t) = \mu$ on $[t_0, t_1]$. Similarly, we prove the distance between two consecutive zeros t_1 and t_2 with u negative on $]t_1, t_2[$ is such that $t_2 - t_1 \geq \frac{\pi}{\sqrt{\nu}}$ with equality if and only if $q(t) = \nu$ on $[t_1, t_2]$. It follows that

$$2\pi \geq t_2 - t_0 \geq \frac{\pi}{\sqrt{\mu}} + \frac{\pi}{\sqrt{\nu}} = 2\pi.$$

This implies $t_2 - t_0 = 2\pi$, $p(t) = \mu$ on $[t_0, t_1]$ and $q(t) = \nu$ on $[t_1, t_2]$ which contradicts (5.8).   ∎

From this Lemma, it is easy to deduce the following admissibility result.

Proposition 5.4 *Let a, b, c, $d \in L^1(0, 2\pi)$ be such that for some $(\mu, \nu) \in F_2$ and some set $I \subset [0, 2\pi]$ of positive measure,*

$$a(t) \leq b(t) \leq \mu, \, c(t) \leq d(t) \leq \nu \quad \text{for a.e. } t \in [0, 2\pi],$$
$$b(t) < \mu, \qquad\qquad d(t) < \nu \quad \text{for a.e. } t \in I,$$

Assume moreover that there exists (μ_0, ν_0) in

$$\{(\mu, \nu) \mid \mu > 0, \nu > 0, \frac{1}{\sqrt{\mu}} + \frac{1}{\sqrt{\nu}} > 2\},$$

such that
$$a(t) \leq \mu_0 \leq b(t), c(t) \leq \nu_0 \leq d(t) \quad \text{for a.e. } t \in [0, 2\pi].$$

Then the set $[a, b] \times [c, d]$ is admissible.

Proof : Part (A) of the definition of admissible sets follows from Lemma 5.3 and part (B) from the definition of the Fučík spectrum.   ∎

We are now ready to state and prove the main result of this section.

Theorem 5.5 *Let $f(t, u)$ be a L^1- Carathéodory function such that for some a, b, c, $d \in L^1(0, 2\pi)$,*
$$a(t) \leq \liminf_{u \to +\infty} \frac{f(t, u)}{u} \leq \limsup_{u \to +\infty} \frac{f(t, u)}{u} \leq b(t),$$
$$c(t) \leq \liminf_{u \to -\infty} \frac{f(t, u)}{u} \leq \limsup_{u \to -\infty} \frac{f(t, u)}{u} \leq d(t),$$

uniformly in t.

Assume the set $[a, b] \times [c, d]$ is admissible and there exist $W^{2,1}$-lower and upper solutions $\alpha(t)$ and $\beta(t)$ such that $\alpha(t) > \beta(t)$ on $[0, 2\pi]$.

Then, the problem (5.1) has at least one solution u such that for some $t_0 \in [0, 2\pi[$,

$$\beta(t_0) \leq u(t_0) \leq \alpha(t_0).$$

Proof : The proof uses degree theory and proceeds in several steps. Throughout, we assume that α and β are lower and upper solutions which are not already solutions of problem (5.1).

Step 1 - Construction of a homotopy. Define the functions

$$
\begin{aligned}
\gamma(t, u, \lambda) &:= u, && \text{if } u \notin [\beta(t) - 2\lambda, \beta(t)] \cup [\alpha(t), \alpha(t) + 2\lambda], \\
&:= 2u, && \text{if } u \in [\beta(t) - 2\lambda, \beta(t) - \lambda] \cup [\alpha(t) + \lambda, \alpha(t) + 2\lambda], \\
&:= \alpha(t), && \text{if } u \in [\alpha(t), \alpha(t) + \lambda], \\
&:= \beta(t), && \text{if } u \in [\beta(t) - \lambda, \beta(t)], \\
f_\lambda(t, u) &:= f(t, \gamma(t, u, \lambda)), \\
k(t, u) &:= f(t, \alpha(t)) + p^1(t)(u - \alpha(t)), && \text{if } u \geq \alpha(t), \\
&:= f(t, u), && \text{if } \beta(t) < u < \alpha(t), \\
&:= f(t, \beta(t)) + q^1(t)(u - \beta(t)), && \text{if } u \leq \beta(t),
\end{aligned}
$$

and consider the homotopy

$$
\begin{aligned}
&\ddot{u} + (1 - \lambda)f_\lambda(t, u) + \lambda k(t, u) = 0, \\
&u(0) = u(2\pi), \ \dot{u}(0) = \dot{u}(2\pi).
\end{aligned}
\tag{5.9}
$$

Define now the compact solution operator

$$
K : L^1(0, 2\pi) \to W^{2,1}(0, 2\pi)
$$

such that for any $f \in L^1(0, 2\pi)$, $u = Kf$ is the unique solution of

$$
\begin{aligned}
&\ddot{u} - u + f = 0, \\
&u(0) = u(2\pi), \ \dot{u}(0) = \dot{u}(2\pi).
\end{aligned}
$$

Consider the space $X := \{u \in C^1([0, 2\pi]) \mid u(0) = u(2\pi)\}$ equipped with the norm $\| \cdot \|_{C^1}$ and the continuous operator $N_\lambda : X \to L^1(0, 2\pi)$ defined by

$$
(N_\lambda u)(t) = (1 - \lambda)f_\lambda(t, u(t)) + \lambda k(t, u(t)) + u(t).
$$

Solutions of (5.9) are the fixed points of the completely continuous operator

$$
T_\lambda : X \to X, u \mapsto T_\lambda u = KN_\lambda u.
$$

At last, for some $K > 0$, we consider the set

$$
\Omega = \{u \in X \mid \exists t_0 \in [0, 2\pi], \beta(t_0) < u(t_0) < \alpha(t_0) \text{ and } \|u\|_{C^1} < K\}.
$$

Step 2 - A priori bounds on the solutions.

Claim - There exists $K > 0$ such that, for any $u \in X$, solution of (5.9), satisfying $\beta(t_0) < u(t_0) < \alpha(t_0)$ for some $t_0 \in [0, 2\pi]$, we have
$$
\|u\|_{C^1} < K.
$$

Let us assume there exist sequences $(u_k)_k \subset X$, $(\lambda_k)_k \subset [0,1]$ and $(t_k)_k \subset [0, 2\pi]$ so that:

(i) $\|u_k\|_{C^1} \geq k$;

(ii) $\lambda = \lambda_k$ and $u = u_k$ satisfy (5.9);

(iii) $\beta(t_k) < u_k(t_k) < \alpha(t_k)$.   (5.10)

Let $r > \max(\|\alpha\|_\infty, \|\beta\|_\infty) + 2$ and define

$$P(t,u) := \max(a(t), \min(\tfrac{f(t,u)}{u}, b(t))), \qquad \text{if } u \geq r,$$
$$:= P(t,r) \qquad\qquad\qquad\qquad\qquad \text{if } u < r,$$
$$Q(t,u) := \max(c(t), \min(\tfrac{f(t,u)}{u}, d(t))), \qquad \text{if } u \leq -r,$$
$$:= Q(t,-r) \qquad\qquad\qquad\qquad\qquad \text{if } u > -r,$$
$$h_\lambda(t,u) := f_\lambda(t,u) - P(t,u)u^+ + Q(t,u)u^-.$$

Notice that

$$P(t,u) \in [a(t), b(t)], \quad Q(t,u) \in [c(t), d(t)]$$

and $h_\lambda(t,u)$ is such that for any $\epsilon > 0$ there exists $\gamma_1 \in L^1(0, 2\pi)$ so that for all $u \in \mathbb{R}$ and a.e. $t \in [0, 2\pi]$

$$|h_\lambda(t,u)| \leq \epsilon|u| + \gamma_1(t).$$

In a similar way, we can write

$$k(t,u) = p^1(t)u^+ - q^1(t)u^- + r(t,u),$$

where $r(t,u)$ is such that for some $\gamma_2 \in L^1(0, 2\pi)$ for all $u \in \mathbb{R}$ and a.e. $t \in [0, 2\pi]$

$$|r(t,u)| \leq \gamma_2(t).$$

Next, we verify that the functions

$$v_k(t) = \frac{u_k(t)}{\|u_k(t)\|_{C^1}}$$

are such that

$$\ddot{v}_k + \hat{p}_k(t)v_k^+ - \hat{q}_k(t)v_k^- + \hat{s}_k(t) = 0,$$
$$v_k(0) = v_k(2\pi), \ \dot{v}_k(0) = \dot{v}_k(2\pi),$$

where

$$\hat{p}_k(t) = (1 - \lambda_k)P(t, u_k(t)) + \lambda_k p^1(t), \quad \hat{p}_k(t) \in [a(t), b(t)],$$
$$\hat{q}_k(t) = (1 - \lambda_k)Q(t, u_k(t)) + \lambda_k q^1(t), \quad \hat{q}_k(t) \in [c(t), d(t)]$$

and

$$\hat{s}_k(t) = (1 - \lambda_k)\frac{h_{\lambda_k}(t, u_k(t))}{\|u_k(t)\|_{C^1}} + \lambda_k \frac{r(t, u_k(t))}{\|u_k(t)\|_{C^1}}.$$

The functions $\hat{s}_k(t)$ are such that for any $\epsilon > 0$, there exists $\gamma_3 \in L^1(0, 2\pi)$ so that for any k

$$|\hat{s}_k(t)| \leq \epsilon + \frac{\gamma_3(t)}{\|u_k(t)\|_{C^1}}.$$

Hence, if k is large enough, $\|\hat{s}_k(t)\|_{L^1} \leq 2\epsilon$, i.e. $\hat{s}_k \xrightarrow{L^1} 0$. From the equation

$$\ddot{v}_k + \hat{p}_k(t)v_k^+ - \hat{q}_k(t)v_k^- + \hat{s}_k(t) = 0,$$

it is clear that the functions $\ddot{v}_k$ are uniformly bounded by a function in $L^1(0, 2\pi)$. Hence, going to a subsequence, we can assume $v_k \xrightarrow{c^1} v$ and, using Dunford-Pettis Theorem (see [20]), $\hat{p}_k \xrightarrow{L^1} \hat{p} \in [a, b]$ and $\hat{q}_k \xrightarrow{L^1} \hat{q} \in [c, d]$. From the closeness of the derivative, we deduce that $v \in W^{2,1}(0, 2\pi)$ and

$$\ddot{v} + \hat{p}(t)v^+ - \hat{q}(t)v^- = 0,$$
$$v(0) = v(2\pi), \ \dot{v}(0) = \dot{v}(2\pi).$$

As the set $[a, b] \times [c, d]$ is admissible, we also know that the function v is one-signed. On the other hand, we deduce from (4.10) that for some subsequence

$$t_k \to t_0, \quad v_k(t_k) = \frac{u_k(t_k)}{\|u_k\|_{c^1}} \to 0,$$

which implies v is not one-signed.

Step 3 – Solutions of (5.9) for $\lambda \geq 0$ are not on $\partial\Omega$.

Claim – Any solution u of (5.9) with $\lambda \geq 0$ and such that $u \in \overline{\Omega}$, satisfies $u(t_0) < \alpha(t_0)$ for some $t_0 \in [0, 2\pi]$.

Let $u \in \overline{\Omega}$ be a solution of (5.9) with $\lambda \geq 0$ and such that for all $t \in [0, 2\pi]$, $u(t) \geq \alpha(t)$. Hence, $u \in \partial\Omega$ and there exists $t_0 \in [0, 2\pi]$ such that $u(t_0) - \alpha(t_0) = \min_t(u - \alpha) = 0$. Extend α and u by periodicity to all of $\mathbb{R}$. By the definition of a $W^{2,1}$-lower solution, we can find a neighbourhood I_0 of t_0 and $t_1 \in I_0$ such that $\alpha \in W^{2,1}(I_0)$, $(u'(t_1) - \alpha'(t_1))\text{sgn}(t_1 - t_0) > 0$, $u'(t_0) - \alpha'(t_0) = 0$ and a.e. on I_0

$$\ddot{u}(t) - \ddot{\alpha}(t) = -(f(t, \alpha(t)) + \ddot{\alpha}(t)) - \lambda p^1(t)(u(t) - \alpha(t)) \leq 0,$$

which gives a contradiction.

In the same way we prove the following claim.

Claim – Any solution u of (5.9) with $\lambda \geq 0$ and such that $u \in \overline{\Omega}$, satisfies $u(t_0) > \beta(t_0)$ for some $t_0 \in [0, 2\pi]$.

Step 4 – Computation of the degree.

From Steps 2 and 3, it is clear that the degree

$$\deg(I - KN_\lambda, \Omega)$$

is defined and independent of $\lambda \in [0, 1]$. For $\lambda = 1$, (5.9) reduces to

$$\ddot{u} + k(t, u) = 0,$$
$$u(0) = u(2\pi), \ \dot{u}(0) = \dot{u}(2\pi). \tag{5.11}$$

Claim – *There is no solution u of* (5.11) *such that for all $t \in [0, 2\pi]$,*
$$u(t) \geq \alpha(t).$$

By contradiction, assume $u \geq \alpha$ on $[0, 2\pi]$. Let $\hat{\alpha}$ be the solution of

$$\ddot{u} - u + f(t, \alpha(t)) + \alpha(t) = 0,$$
$$u(0) = u(2\pi), \quad \dot{u}(0) = \dot{u}(2\pi).$$

Observe that $\hat{\alpha} \geq \alpha$. If not, $\min_t(\hat{\alpha}(t) - \alpha(t)) = \hat{\alpha}(t_0) - \alpha(t_0) < 0$ and by definition of a $W^{2,1}$-lower solution, we have a neighbourhood I_0 of t_0 such that $\alpha \in W^{2,1}(I_0)$ and for a.e. $t \in I_0$, $\hat{\alpha}(t) < \alpha(t)$,

$$\ddot{\hat{\alpha}}(t) - \ddot{\alpha}(t) = -f(t, \alpha(t)) - \ddot{\alpha}(t) + \hat{\alpha}(t) - \alpha(t) < 0,$$

which contradicts the definition of t_0. Moreover, as α is not a solution, $\hat{\alpha} \not\equiv \alpha$. Observe now that $v = u - \hat{\alpha}$ satisfies

$$\ddot{v} = -p^1(t)(u - \alpha(t)) - (\hat{\alpha}(t) - \alpha(t)),$$
$$v(0) = v(2\pi), \quad \dot{v}(0) = \dot{v}(2\pi).$$

Hence, we have the contradiction

$$0 = \dot{v}(2\pi) - \dot{v}(0) = -\int_0^{2\pi} [p^1(t)(u(t) - \alpha(t)) + (\hat{\alpha}(t) - \alpha(t))] \, dt < 0.$$

In a similar way we prove the following.

Claim – *There is no solution u of* (5.11) *such that for all $t \in [0, 2\pi]$,*
$$u(t) \leq \beta(t).$$

As a consequence of the above claims, we have

$$\deg(I - KN_1, \Omega) = \deg(I - KN_1, B_R),$$

where B_R is a ball of radius large enough so that $B_R \supset \Omega$.

Define now for $\mu \in [0, 1]$, the functions

$$
\begin{aligned}
k_\mu(t, u) &:= f(t, \alpha(t)) + p^\mu(t)(u - \alpha(t)), && \text{if } u \geq \alpha(t), \\
&:= f(t, u), && \text{if } \beta(t) < u < \alpha(t), \\
&:= f(t, \beta(t)) + q^\mu(t)(u - \beta(t)), && \text{if } u \leq \beta(t),
\end{aligned}
$$

and the operators $\hat{N}_\mu : X \to L^1(0, 2\pi)$, where

$$(\hat{N}_\mu u)(t) = \mu k_\mu(t, u) + (1 - \mu)(p^\mu u^+ - q^\mu u^-) + u.$$

Next, we consider the problems

$$\ddot{u} + p^\mu u^+ - q^\mu u^- + \mu(k_\mu(t, u) - p^\mu u^+ + q^\mu u^-) = 0,$$
$$u(0) = u(2\pi), \quad \dot{u}(0) = \dot{u}(2\pi). \tag{5.12}$$

The functions $k_\mu(t, u) - p^\mu u^+ + q^\mu u^-$ are uniformly bounded in $L^1(0, 2\pi)$. Hence, using the argument of Step 2, we can find a priori bounds on the solutions of (5.12) and, if R is large enough,

$$\deg(I - K\hat{N}_0, B_R) = \deg(I - K\hat{N}_1, B_R) = \deg(I - KN_1, B_R).$$

As further $I - K\hat{N}_0$ is linear and one-to-one, one has

$$\deg(I - K\hat{N}_0, B_R) = \pm 1,$$

which proves

$$\deg(I - KN_1, \Omega) = \pm 1.$$

The theorem follows. ∎

As a first application, consider the problem

$$\ddot{u} + g(u) + h(t) = 0,$$
$$u(0) = u(2\pi), \quad \dot{u}(0) = \dot{u}(2\pi). \tag{5.13}$$

We can apply the previous results to prove the following proposition which does not impose a control from below (such as in (5.5)) on the nonlinearity.

Proposition 5.6 *Let g be a continuous function, $h \in L^\infty(0, 2\pi)$ and (μ, ν) $\in F_2$. Assume that*

$$b := \limsup_{u \to +\infty} \frac{g(u)}{u} < \mu, \quad d := \limsup_{u \to -\infty} \frac{g(u)}{u} < \nu$$

and there exist $W^{2,1}$-lower and upper solutions $\alpha(t)$ and $\beta(t)$ such that $\alpha(t) > \beta(t)$ on $[0, 2\pi]$.
Then, the problem (5.13) has at least one solution u.

Exercise 5.1 Prove Proposition 5.6.
Hint : Use the fact that in case

$$\liminf_{u \to +\infty} g(u) = -\infty \quad \text{or} \quad \limsup_{u \to -\infty} g(u) = +\infty,$$

we can find well-ordered upper and lower solutions.

Notice that Proposition 5.6 does not guarantee that for some $t_0 \in [0, 2\pi]$, $\beta(t_0) \leq u(t_0) \leq \alpha(t_0)$.

As a complement to Proposition 5.6 we can work out conditions which imply the existence of upper and lower solutions.

Proposition 5.7 *Let g be a continuous function, $h \in L^\infty(0, 2\pi)$ and (μ, ν) $\in F_2$. Assume that*

$$0 < b := \limsup_{u \to +\infty} \frac{g(u)}{u} < \mu, \quad 0 < d := \limsup_{u \to -\infty} \frac{g(u)}{u} < \nu.$$

Then, the problem (5.13) has at least one solution u.

Proof : By Proposition 5.6, we just have to prove that there exists lower and upper solutions, α and β, such that $\alpha(t) > \beta(t)$ on $[0, 2\pi]$. Observe that, thanks to the assumptions $b > 0$, $d > 0$, we have $(a_n)_n \subset \mathbb{R}$ and $(b_n)_n \subset \mathbb{R}$ such that

$$a_n \to +\infty \quad \text{and} \quad g(a_n) \to +\infty,$$
$$b_n \to -\infty \quad \text{and} \quad g(b_n) \to -\infty.$$

Hence, as $h \in L^\infty(0, 2\pi)$, we have that, for n large enough, $\alpha(t) := a_n$, $\beta(t) := b_n$ are the required lower and upper solutions. ∎

Upper and lower solutions in the reversed order appear in several applications. For example, we can use Landesman-Lazer conditions.

Proposition 5.8 *Assume $f(t, u)$ is a L^1- Carathéodory function such that for some $f_+ \in L^1(0, 2\pi)$*

$$\liminf_{u \to +\infty} f(t, u) = f_+(t)$$

and

$$\overline{f}_+ := \frac{1}{2\pi} \int_0^{2\pi} f_+(t) \, dt > 0.$$

Then for any $R > 0$, there exists a function $\alpha(t) \geq R$ which is a lower solution for (5.1).

Proof : Let w be the solution of

$$\ddot{w} + f_+(t) - \overline{f}_+ = 0,$$
$$w(0) = w(2\pi), \ \dot{w}(0) = \dot{w}(2\pi).$$

Then for A large enough, $\alpha(t) = w(t) + A$ is a lower solution. ∎

Exercise 5.2 Obtain upper solutions from the argument of Proposition 5.8.

Consider now the problem

$$\ddot{u} + au \sin^2 u^2 + h(t) = 0,$$
$$u(0) = u(2\pi), \ \dot{u}(0) = \dot{u}(2\pi),$$
$$\tag{5.14}$$

where $h \in L^2(0, 2\pi)$ and $a \in \,]0, 1[$. Since h can be unbounded, we cannot find constant upper or lower solutions. However, we can obtain them from the following proposition.

Proposition 5.9 *Let g be a continuous function and $h \in L^2(0, 2\pi)$. Assume that there exist sequences $(a_k)_k$, $(b_k)_k$ and $(c_k)_k$ of positive numbers such that*

(i) $\lim\limits_{k\to\infty} a_k = \lim\limits_{k\to\infty} b_k - c_k = +\infty;$

(ii) for some $\epsilon > 0$ and any k, $a_k c_k \geq \epsilon;$

(iii) for any $u \in [b_k - c_k, b_k + c_k]$, $g(u) \geq a_k$.

Then, for any $R > 0$, there exists a function $\alpha(t) \geq R$, which is a lower solution for (5.13).

Proof : Let $R > 0$ be fixed. Consider the problem

$$\ddot{u} - du + h(t) = 0,$$
$$u(0) = u(2\pi), \quad \dot{u}(0) = \dot{u}(2\pi),$$

where $d > 0$. Any solution u of this problem is such that

$$\int_0^{2\pi} \dot{u}^2 \, dt + d \int_0^{2\pi} u^2 \, dt = \int_0^{2\pi} hu \, dt.$$

Hence, we have the a priori estimates

$$\|u\|_{L^2} \leq \frac{1}{d}\|h\|_{L^2} \quad \text{and} \quad \|\dot{u}\|_{L^2} \leq \frac{1}{\sqrt{d}}\|h\|_{L^2}.$$

Also, for some t_0,

$$|u(t_0)| \leq \frac{1}{d}\|h\|_{L^2}.$$

Hence, we obtain

$$u^2(t) = u^2(t_0) + 2\int_{t_0}^t u(s)\dot{u}(s) \, ds \leq \frac{\|h\|_{L^2}^2}{d^2}(1 + 2d^{\frac{1}{2}}).$$

Let us choose d_0 such that

$$\frac{\|h\|_{L^2}^2}{d_0}(1 + 2d_0^{\frac{1}{2}}) = {'}\epsilon,$$

and k so that

$$a_k \geq \|h\|_{L^2}(1 + 2d_0^{\frac{1}{2}})^{\frac{1}{2}}, \quad b_k - c_k \geq R$$

and next $d \geq d_0$ so that $a_k = \|h\|_{L^2}(1 + 2d^{\frac{1}{2}})^{\frac{1}{2}}$. For this choice of d and k, and all $t \in [0, 2\pi]$, we have

$$a_k \geq d|u(t)|$$

and $'$

$$c_k \geq \frac{\epsilon}{a_k} \geq \|h\|_{L^2}\frac{(1 + 2d^{\frac{1}{2}})^{\frac{1}{2}}}{d} \geq |u(t)|.$$

Now, it is easy to see that $\alpha(t) = b_k + u(t) \geq R$ is a lower solution since

$$\ddot{\alpha} + g(\alpha) + h(t) \geq du - h(t) + a_k + h(t) \geq 0. \qquad \blacksquare$$

Exercise 5.3 Prove that if in Proposition 5.9, (iii) is replaced by

(iii') for any $u \in [-b_k - c_k, -b_k + c_k]$, $g(u) \leq -a_k$,

then, for any $R > 0$, there exists a function $\beta(t) \leq -R$, which is a upper solution for (5.13).

Now we can solve the problem (5.14).

Exercise 5.4 Prove existence of solutions for (5.14).

We can also use the localization of solutions to obtain a multiplicity result.

Theorem 5.10 Let $f : [0, 2\pi] \times \mathbb{R} \to \mathbb{R}$ be a L^1-Carathéodory function. Assume that there are real numbers $\beta_1 \leq \alpha_1 < \alpha_2 \leq \beta_2$ such that, for all $t \in [0, 2\pi]$,

$$f(t, \beta_i) \leq 0 \leq f(t, \alpha_i), \quad \text{for } i = 1, 2,$$
$$f(t, x) \leq 0, \qquad \qquad \text{if } x \leq \beta_1.$$

Then the problem (5.1) has at least two solutions u_1, u_2 with $\inf u_1(t) \leq \alpha_1$ and $\alpha_2 \leq u_2(t) \leq \beta_2$.

Proof : Observe that, for $i = 1, 2$, β_i is an upper solution and α_i is a lower solution. Hence the existence of u_2 follows from Theorem 2.2.

Step 1 - Claim: There exists $K < \beta_1$ such that, every solution u of (5.1) with $\beta_1 \leq \max u(t) \leq \beta_2$ satisfies $u(t) \geq K$ on $[0, 2\pi]$.

Assume there exists $t_0, t_1 \in [0, 2\pi]$ such that $u(t_0) = \sup u(t) \geq \beta_1$ and $u(t_1) = \inf u(t)$. Let $h \in L^1(0, 2\pi)$ be such that, for a.e. $t \in [0, 2\pi]$ and all $u \in [\beta_1, \beta_2]$, $|f(t, u)| \leq h(t)$. Then we have

$$\inf u(t) = u(t_1) > \beta_1 - \int_{t_0}^{t_1} (t_1 - s)h(s)\, ds =: K$$

which proves the claim.

Step 2 - Conclusion

Consider the modified problem

$$\ddot{u} - u + f(t, \gamma(t, u)) + \gamma(t, u) = 0,$$
$$u(0) = u(2\pi), \quad \dot{u}(0) = \dot{u}(2\pi),$$

(5.15)

where

$$\gamma(t, u) = K, \text{ if } u < K,$$
$$= u, \text{ if } K \leq u \leq \beta_2,$$
$$= \beta_2, \text{ if } u > \beta_2.$$

It is easy to observe that the problem (5.15) satisfies the conditions of Theorem 5.5 with $\alpha(t) = \alpha_1$ and $\beta(t) = \beta_1$. Hence there exists a solution u_1 of (5.15) such that, for some $t_0 \in [0, 2\pi]$, $\beta_1 \leq u_1(t_0) \leq \alpha_1$. By the construction of the modified problem, it is now easy to observe that, for all $t \in [0, 2\pi]$, $u_1(t) \leq \beta_2$ and, by Step 1, $u_1(t) \geq K$ on $[0, 2\pi]$. Hence, u_1 is a solution of (5.1) with $\inf u_1(t) \leq \alpha_1$. ∎

Remark 5.1 We can replace the condition on α_1 and α_2 by the following: there exists $\alpha_3 \in [\beta_1, \beta_2]$ which is a strict lower solution.

5.3   Monotone methods

If the function $f(t, u)$ satisfies some monotone assumption, we have seen in Section 1.5 that solutions can be obtained from an iterative process. This can be generalized to some extend to the case the upper and lower solutions are in the reversed order. To achieve this goal, we need the following preliminary result.

Lemma 5.11 *The problem*

$$\ddot{u} + ku = f(t),$$
$$u(0) - u(2\pi) = 0, \ \dot{u}(0) - \dot{u}(2\pi) = \lambda, \tag{5.16}$$

has a positive solution for any $\lambda \geq 0$ and $f \in C([0, 2\pi])$, $f(t) \geq 0$, $f \not\equiv 0$, if and only if $0 < k \leq \frac{1}{4}$.

Proof : Notice first that if $k < 0$, we deduce from Lemma 2.19 that $u \leq 0$. Further, if $k = 0$ and $f \in C([0, 2\pi])$, $f(t) \geq 0$, $f \not\equiv 0$, the problem (5.16) has no solution.

Claim 1 – *If $0 < k \leq \frac{1}{4}$, $\lambda \geq 0$ and $f(t) \geq 0$, $f \not\equiv 0$, solutions $u(t)$ of (5.16) are one-signed.*

Let t_0 be a zero of u. Extend u by periodicity, define $v(t) = \sin\sqrt{k}(t - t_0)$ and compute

$$\dot{u}(t_0) \sin 2\pi\sqrt{k} \geq (\dot{u}v - \dot{v}u)|_{t_0}^{2\pi} + (\dot{u}v - \dot{v}u)|_{2\pi}^{t_0 + 2\pi}$$
$$= \int_{t_0}^{t_0 + 2\pi} f(s) \sin\sqrt{k}(s - t_0)\, ds > 0.$$

If $k = \frac{1}{4}$, this is contradictory. If $k \neq \frac{1}{4}$, this implies $\dot{u}(t_0) > 0$ and u cannot be a periodic function.

Claim 2 – *If $0 < k \leq \frac{1}{4}$, $\lambda \geq 0$ and $f(t) \geq 0$, $f \not\equiv 0$, any one-signed solution of (5.16) is positive.*

Direct integration of (5.16) gives

$$k \int_0^{2\pi} u(s)\, ds = \int_0^{2\pi} f(s)\, ds + \lambda > 0.$$

Hence, we have $u(t) > 0$.

Claim 3 – *If $k > \frac{1}{4}$, there exist nonnegative, nontrivial functions f and $\lambda \geq 0$ such that the corresponding solution u of (5.16) is not one-signed.*

Let a, b be such that

$$\frac{1}{4} < a^2 < k < b^2 \quad \text{and} \quad \frac{1}{a} + \frac{1}{b} = 2.$$

Define

$$f(t) := \frac{k - a^2}{a} \sin at, \qquad\qquad \text{if } t \in [0, \tfrac{\pi}{a}],$$

$$:= \frac{k - b^2}{b} \sin b(t - \tfrac{\pi}{a} + \tfrac{\pi}{b}), \qquad \text{if } t \in \left]\tfrac{\pi}{a}, 2\pi\right],$$

and take $\lambda = 0$. Then the function

$$u(t) := \tfrac{1}{a} \sin at, \qquad\qquad \text{if } t \in [0, \tfrac{\pi}{a}],$$

$$:= \tfrac{1}{b} \sin b(t - \tfrac{\pi}{a} + \tfrac{\pi}{b}), \quad \text{if } t \in]\tfrac{\pi}{a}, 2\pi]$$

is a solution of (5.16) which is not one-signed. ∎

We can now state and prove the main result of this section.

Theorem 5.12 *Let $f(t, u)$ be a continuous function and let α and $\beta \leq \alpha$ be lower and upper solutions. Assume that for any $t \in [0, 2\pi]$, and u, v, with $\beta(t) \leq u \leq v \leq \alpha(t)$, we have*

$$\frac{1}{4}u - f(t, u) \leq \frac{1}{4}v - f(t, v).$$

Then, there exist two monotone sequences $(\alpha_n)_n$, $(\beta_n)_n$,

$$\alpha = \alpha_0 \geq \ldots \geq \alpha_{n-1} \geq \alpha_n \geq \ldots \geq \beta_n \geq \beta_{n-1} \geq \ldots \geq \beta_0 = \beta,$$

which converge uniformly to extremal solutions of (5.1) in $[\beta, \alpha]$.

Proof : Let $K : C([0, 2\pi]) \rightarrow C([0, 2\pi])$ be the compact solution operator such that $u = Kf$ is the solution of (5.16) with $k = \tfrac{1}{4}$ and $\lambda = 0$. Let $N : C([0, 2\pi]) \rightarrow C([0, 2\pi])$ be the continuous operator defined by $(Nu)(t) = \tfrac{1}{4}u(t) - f(t, u(t))$. Then the operator KN is completely continuous, increasing, and the proof follows from the argument of Theorem 2.20. ∎

6 Historical and Bibliographical Notes

6.1 The method of upper and lower solutions

In 1931, G. Scorza Dragoni [112] has considered the boundary value problem

$$\ddot{u} + f(t, u, \dot{u}) = 0,$$
$$u(a) = A, \ u(b) = B, \tag{6.1}$$

under the following assumptions:

(i) there exist $\alpha, \beta \in C^2([a, b])$ such that $\alpha(t) \leq \beta(t)$ on $[a, b]$;
(ii) f is defined, continuous and bounded on

$$E = \{(t, x, y) \in [a, b] \times \mathbb{R}^2 \mid \alpha(t) \leq x \leq \beta(t)\};$$

(iii) α and β are solutions of the differential equation

$$\ddot{u} + f(t, u, \dot{u}) = 0;$$

(iv) $\alpha(a) \leq A \leq \beta(a)$, $\alpha(b) \leq B \leq \beta(b)$;

(v) $f(t, \alpha(t), .)$ and $f(t, \beta(t), .)$ are monotone for all $t \in [a, b]$.

His method of proof is already two steps. First, he introduces a modified problem

$$\ddot{u} + \overline{f}(t, u, \dot{u}) = 0,$$
$$u(a) = A, \ u(b) = B,$$

where $\overline{f}$ is bounded, equal to f on E and decreasing for $u < \alpha(t)$ and $u > \beta(t)$. For this modified problem, he obtains then existence of a solution u. Second, he proves, using a maximum principle type argument that $\alpha \leq u \leq \beta$.

A first extension concerns the dependence of the nonlinearity in $\dot{u}$. The same year, G. Scorza Dragoni [113] extends already his result to functions α and β which satisfy differential inequalities. A major step however is due to M. Nagumo [98] in 1937 who introduces the so-called Nagumo condition which provides a somewhat final approach of the $\dot{u}$ dependence. This paper was further generalized by G. Scorza Dragoni [114]. Let us notice also that in 1963 H. Knobloch [76] extended the result of M. Nagumo to the periodic problem.

Lower and upper solutions α and $\beta \in C^2([a, b])$ are classical notions that can be found in several textbooks such as P. B. Bailey - L. F. Shampine - P. E. Waltman [13], S. Fučík [52], L. C. Piccinini - G. Stampacchia - G. Vidossich [105], N. Rouche - J. Mawhin [106].

The idea to introduce corner in the lower and upper solutions might go back to M. Nagumo [98]. In this paper, he considers functions α and $\beta \in C^1$ so that $\dot{\alpha}$ and $\dot{\beta}$ have right and left derivatives. A major step, however, is the notion of quasi-subsolution coined by M. Nagumo [99]. A first order condition, similar to $D_- \alpha(t_0) < D^+ \alpha(t_0)$, appears in H. Knobloch [76]. Similar notions of lower and upper solutions can be found in L. K. Jackson [70], K. Schmitt [109], P. Habets - M. Laloy [60], C. Fabry - P. Habets [48].

G. Scorza Dragoni will extend his results in [115] and [116] to the case where f is a L^1-Carathéodory function. In this case, the notion of lower and upper solution is generalized by assuming that $\alpha, \beta \in C^1$ and the functions

$$-\dot{\alpha}(t) - \int_a^t f(s, \alpha(s), \dot{\alpha}(s)) \, ds \text{ and } \dot{\beta}(t) + \int_a^t f(s, \beta(s), \dot{\beta}(s)) \, ds$$

are nonincreasing on $[a, b]$. The use of lower and upper solutions to investigate $W^{2,1}$-solutions also appears in H. Epheser [45]. More recently, one can quote K. Ako [4], V. V. Gudkov - A. Ja. Lepin [58], A. Adje [1], [2], P. Habets - L. Sanchez [62]. These papers present a variety of definitions which are strongly related. The generalization of the L^1-Carathéodory condition as considered in Theorem 2.6 is based on P. Habets - F. Zanolin [63], [64], where the Dirichlet boundary value problem is investigated with an other definition of lower and upper solutions.

We can find results on the existence of periodic solutions for a problem with singular forces in A.C. Lazer - S. Solimini [82] and P. Habets - L. Sanchez [62].

The use of lower and upper solutions in the study of resonance and non-resonance seems to go back to J. L. Kazdan - F. W. Warner [74]. Theorem 2.10 can be found in J. Mawhin [89].

The idea to associate a degree to a pair of strict lower and upper solutions is due to H. Amann [7], [8]. Until recently, this kind of results was only made for the continuous case.

We found in J. Mawhin [89] the particular case of Theorem 2.13 when $f : [0, 2\pi] \times \mathbb{R} \to \mathbb{R}$ is continuous. In C. De Coster [34], C. De Coster - M.R. Grossinho - P. Habets [35], P. Habets - P. Omari [61] and C. De Coster - P. Habets [36], we find a study of the Carathéodory case for a Dirichlet problem with f independent of $\dot{u}$.

It seems that the abstract idea used in Proposition 2.14 — two pairs of lower and upper solutions with an ordering condition implies the existence of three solutions — goes back to Y.S. Kolesov [77] and was extended by H. Amann [6]. The first one who proved it with degree theory seems to be H. Amann [7], [8] (see also [9]). Extensions are also given by F. Inkmann [69], R. Shivaji [118] and Bongsoo Ko [19]. The results presented here improve all these in the special case of ODE and extend them to the Carathéodory case.

The idea of Theorem 2.15 is due to K.J. Brown - H. Budin [23]. A more precise study of it has been made in C. De Coster [34] for the p-Laplacian with Dirichlet boundary conditions and f a Carathéodory function independent of $\dot{u}$. We can find Theorem 2.16 in C. De Coster - M.R. Grossinho - P. Habets [35].

An early result using lower and upper techniques in the study of a pendulum equation can be found in H. W. Knobloch [76]. He obtains the existence of the first solution in Proposition 2.14. The existence of the second solution is due to J. Mawhin - M. Willem [96]. We refer the interested reader to J. Mawhin [89], J. Mawhin - M. Willem [96] and J. Mawhin [87] for the history of this problem. See also G. Fournier - J. Mawhin [51], R. Kannan - R. Ortega [73] and J. Mawhin [91] for more results in this direction. For an history of the pendulum equation from Galileo to recent results we refer to J. Mawhin [92].

For what concerns the variational methods, it seems that the first ones who have proved that between a lower and an upper solution we have a minimum of the related functional are independently K.C. Chang [25], [26] and D.G. de Figueiredo - S. Solimini [42] (see also J. Mawhin [90] and H. Brezis - L. Nirenberg [21]). The application presented here is a particular case of a result of P. Omari - F. Zanolin [104].

The story of the iterative method seems to be difficult to reconstruct. The iteration method of Theorem 2.20 can already be found in the book of R. Courant and D. Hilbert [31] (vol. 2, pp 367-374). It seems that a result like Theorem 2.20 for Dirichlet boundary conditions was first proved by S.I. Hudjaev [67] with a different method. Without being aware of this research, L.F. Shampine - G.M. Wing [117], D. Sattinger [107], [108] and H. Amann [6] rediscovered Theorem 2.20 in the Dirichlet case. All of these papers were inspired by the basic paper of H.B. Keller - D.S. Cohen [75]. For the periodic problem, it seems that we have to wait S. Leela [83] and the first one who considers this method with Carathéodory functions seems to be J.J. Nieto [100].

To replace the upper and lower solutions within the general context of boundary value problems for nonlinear ordinary differential equations, the reader can consult J. Mawhin [93].

6.2   Systems with singularities

The Emden-Fowler equation (3.2) has been investigated by S.D. Taliaferro [119]. Contributions to the more general problem (3.1) were obtained by several authors [14], [18], [53], [59] [121] but most of their results rely on the fact that, $f(t, u)$ being positive, the solutions are concave. This assumption was given up in J. Janus and J. Myjak [71] for the equation

$$\ddot{u} + \frac{f(t)}{u^\sigma} = h(t)$$

and for the general case in P. Habets and F. Zanolin [63] and [64].

Theorem 3.1 can be found in [64]. We can find other related results in A. G. Lomtatidze [85].

The sublinear case i.e. the case where

$$\lim_{u \to 0} \frac{f(t, u)}{u} = +\infty, \quad \lim_{u \to +\infty} \frac{f(t, u)}{u} = 0,$$

is classically considered by lower and upper solutions when there are no singularity. See for example D.G. de Figueiredo [38], D.G. Costa - J.V.A. Goncalves [30].

It is easy to observe that in the sub-superlinear case, i.e. in the case $\lim_{u \to 0} \frac{f(t,u)}{u} = +\infty$, $\lim_{u \to +\infty} \frac{f(t,u)}{u} = +\infty$, the assumptions at 0 and at $+\infty$ alone do not give the existence of positive solutions. To obtain such a result for an elliptic Dirichlet problem, D.G. de Figueiredo - P.L. Lions [40] assume moreover the existence of a strict upper solution and prove the existence of two positive solutions. An alternative assumption is introduced in F.J.S.A. Corrêa [29]. In the particular case of an ODE, the condition of F.J.S.A. Corrêa implies the existence of a strict upper solution.

For more informations, see P.L. Lions [84], D.G. de Figueiredo [38], [39] and the references therein.

Section 3.2 comes from C. De Coster, M. R. Grossinho and P. Habets [35].

6.3   An Ambrosetti-Prodi problem

The study of the Ambrosetti-Prodi problem has started with the paper of A. Ambrosetti - G. Prodi [12] who consider the problem

$$\Delta u + f(u) = h(x), \text{ in } \Omega,$$
$$u = 0, \text{ on } \partial\Omega,$$

where f is a convex function which satisfies

$$0 < \lim_{u \to -\infty} f'(u) < \lambda_1 < \lim_{u \to +\infty} f'(u) < \lambda_2$$

where λ_1, λ_2 are the two first eigenvalues of the corresponding eigenvalue problem. They prove that in the space $H = \mathrm{Im}(\Delta u + f(u))$, there is a manifold M which separates the space in two regions O_0 and O_2 such that the above problem has zero, exactly one or exactly two solutions according to $h \in O_0$, $h \in M$ or $h \in O_2$. In 1975, M.S. Berger - E. Podolak [17] use the decomposition of h in terms of the first eigenfunction φ of $-\Delta$ with Dirichlet condition i.e. $h(x) = \nu \varphi(x) + \tilde{h}(x)$, where $\int_\Omega \tilde{h}(x) \varphi(x)\, dx = 0$, and characterize the manifold M in terms of the parameter ν. In the same year, J. L. Kazdan - F. W. Warner [74] have used the lower and upper solution method to study this problem but were only able to prove the existence of one solution if $h \in O_2$. It is only in 1978 that E.N. Dancer [32] and H. Amann - P. Hess [11] have obtained independently the multiplicity result by combining lower and upper solutions technique with degree theory. An exact count of solutions was then obtained by H. Berestycki [16] in case f is convex. See also the survey paper of D.G. de Figueiredo [37]. Later, R. Chiappinelli - J. Mawhin - R. Nugari [27] have considered the problem

$$\ddot{u} + u + f(t,u) = \nu \varphi(t),$$
$$u(0) = 0, \ u(\pi) = 0,$$
$$(6.2)$$

where $\varphi(t) = \sqrt{\frac{2}{\pi}} \sin t$. They have proved that, if $\lim_{|u| \to +\infty} f(t,u) = +\infty$, uniformly in t, then there exist ν_0 and $\nu_1 > \nu_0$ such that

(i) if $\nu < \nu_0$, the problem (6.2) has no solution;
(ii) if $\nu_0 \leq \nu \leq \nu_1$, the problem (6.2) has at least one solution;
(iii) if $\nu_1 < \nu$, the problem (6.2) has at least two solutions.

The question is then to know whether $\nu_0 = \nu_1$. Under the additional condition that $f(t,u) + Ku$ is nondecreasing on $[-u_0, u_0]$, they prove that $\nu_0 = \nu_1$. C. Fabry [46], using another argument, proves $\nu_0 = \nu_1$ with the help of a Hölder condition. R. Chiappinelli - J. Mawhin - R. Nugari [27] have also considered a generalized case where instead of the coercitivity condition they assume some growth restriction related to the classical Landesman-Lazer condition.

The results presented here are due to C. De Coster and P. Habets [36]. Other related results can be obtained such as those presented in Problems 4.1, 4.2, 4.3, 4.4. For these kind of results we refer the interested reader to H. Berestycki [16], A.C. Lazer and P.J. McKenna [79], [80], [81] and G. Harris [65].

6.4 Non-ordered upper and lower solutions

Solvability of boundary value problems in case α and β do not satisfy the "usual" ordering condition

$$\alpha(t) \leq \beta(t) \ \text{ for all } t \in [0, 2\pi]$$

was explicitly raised in the early seventies by D. Sattinger [107] and became the subject of some works in the last two decades [111], [10], [101], [54], [55], [56], [57], [61].

In [10], one of the first contribution to this problem, H. Amann, A. Ambrosetti and G. Mancini proved that for a PDE problem the existence of a lower solution α and of an upper solution β (satisfying no ordering condition) implies the solvability of (5.1), provided that $f(t, u)$ is bounded. Theorem 5.1 adapts this result to the periodic problem (5.1). More recently, the above mentioned result was generalized in J.P. Gossez and P. Omari [57] to selfadjoint problems with unbounded nonlinearity.

Theorem 5.5 is based on P. Habets and P. Omari [61] where nonselfadjoint problems are considered with ordered upper and lower solutions, i.e. satisfying (5.2).

Monotone methods for nonordered upper and lower solutions is based on an anti-maximum principle [66], [28]. Results in this direction are due to P. Omari and M. Trombetta [102] and A. Cabada and L. Sanchez [24].

The classical Landesman-Lazer conditions as in Proposition 5.8 apply to a boundary value problem which is asymptotically resonant. This type of result is due to E.M. Landesman and A.C. Lazer [78]. Theorem 5.2 is due to D.G. de Figueiredo and W.N. Ni [41] and has been extended to some class of unbounded nonlinearities by R. Iannacci, M.N. Nkashama and J.R. Ward [68]. A proof of this extension by lower and upper solutions can be found in J.P. Gossez and P. Omari [56]. Proposition 5.9 is a variant of a result of R. Kannan and K. Nagle [72]. Theorem 5.10 improves a result of A. Tineo [120].

For other related results we refer to A. Fonda [49] and P. Omari and W. Ye [103].

References

[1] A. Adje, *Sur et sous solutions dans les équations différentielles discontinues avec conditions aux limites non linéaires*, Thèse de Doctorat, U.C.L., Louvain-la-Neuve (1987).

[2] A. Adje, *Sur et sous-solutions généralisées et problèmes aux limites du second ordre*, Bull. Soc. Math. Belg. **42** ser. B (1990), 347-368.

[3] K. Ako, *On the Dirichlet problem for quasi-linear elliptic differential equations of the second order*, J. Math. Soc. Japan **13** (1961), 45-62.

[4] K. Ako, *Subfunctions for ordinary differential equations II*, Funkcialaj Ekvacioj **10** (1967), 145-162.

[5] K. Ako, *Subfunctions for ordinary differential equations VI*, J. Fac. Sci. Univ. Tokyo **16** (1969), 149-156.

[6] H. Amann, *On the existence of positive solutions of nonlinear elliptic boundary value problems*, Indiana Univ. Math. J. **21** (1971), 125-146.

[7] H. Amann, *Existence of multiple solutions for nonlinear elliptic boundary value problems*, Indiana Univ. Math. J. **21** (1972), 925-935.

[8] H. Amann, *On the number of solutions of nonlinear equations in ordered Banach spaces*, J. Funct. Anal. **11** (1972), 346-384.

[9] H. Amann, *Fixed point equations and nonlinear eigenvalue problems in ordered Banach spaces*, SIAM Review **18** (1976), 620-709.

[10] H. Amann, A. Ambrosetti and G. Mancini, *Elliptic equations with noninvertible Fredholm linear part and bounded nonlinearities*, Math. Z. **158** (1978), 179-194.

[11] H. Amann and P. Hess, *A multiplicity result for a class of elliptic boundary value problems*, Proc. Royal Soc. Edinburgh **84A** (1979), 145-151.

[12] A. Ambrosetti and G. Prodi, *On the inversion of some differentiable mappings with singularities between Banach spaces*, Ann. Mat. Pure Appl. **93** (1972), 231-247.

[13] P.B. Bailey, L.F. Shampine and P.E. Waltman, *Nonlinear two point boundary value problems*, Academic Press, New York (1968).

[14] J.V. Baxley, *A singular nonlinear boundary value problem: membrane response of a spherical cap*, SIAM J. Appl. Math. **48** (1988), 497-505.

[15] A.K. Ben-Naoum and C. De Coster, *On the existence and multiplicity of positive solutions of the p-Laplacian separated boundary value problem*, Recherches de mathématique 46, preprint UCL 1995.

[16] H. Berestycki, *Le nombre de solutions de certains problèmes semilinéaires elliptiques*, J. Funct. Anal. **40** (1981), 1-29.

[17] M.S. Berger and E. Podolak, *On the solutions of a nonlinear Dirichlet problem*, Indiana Univ. Math. J. **24** (1975), 837-846.

[18] L.E. Bobisud, D. O'Regan and W.D. Royalty, *Solvability of some nonlinear boundary value problem*, Nonlinear Anal. T.M.A. **12** (1988), 855-869.

[19] Bongsoo Ko, *The third solution of semilinear elliptic boundary value problems and applications to singular perturbation problems*, J. Diff. Equ. **101** (1993), 1-14.

[20] H. Brezis, *Analyse fonctionnelle : Théorie et applications*, Masson, Paris (1983).

[21] H. Brezis and L. Nirenberg, H^1 versus C^1 local minimizers, C. R. Acad. Sci. Paris 317 (1993), 465-472.

[22] R.F. Brown, A topological introduction to nonlinear analysis, Birkhäuser, Boston 1993.

[23] K.J. Brown and H. Budin, Multiple positive solutions for a class of nonlinear boundary value problem, J. Math. Anal. Appl. 60 (1977), 329-338.

[24] A. Cabada and L. Sanchez, A positive operator approach to the Neumann problem for a second order ordinary differential equation, preprint.

[25] K.C. Chang, A variant mountain pass lemma, Scientia Sinica (Series A) 26 (1983), 1241-1255.

[26] K.C. Chang, Variational methods and sub- and super-solutions, Scientia Sinica (Series A) 26 (1983), 1256-1265.

[27] R. Chiappinelli, J. Mawhin and R. Nugari, Generalized Ambrosetti-Prodi conditions for nonlinear two-points boundary value problems, J. Diff. Equ. 69 (1987), 422-434.

[28] P. Clement and L.A. Peletier, An anti-maximum principle for second order elliptic operators, J. Diff. Equ. 34 (1979), 218-229.

[29] F.J.S.A. Corrêa, On pairs of positive solutions for a class of sub-superlinear elliptic problems, Diff. Int. Equ. 5 (1992), 387-392.

[30] D.G. Costa and J.V.A. Goncalves, On the existence of positive solutions for a class of non-selfadjoint elliptic boundary value problems, Applicable Analysis 31 (1989), 309-320.

[31] R. Courant and D. Hilbert, Methods of Mathematical Physics, vol. II, Interscience, New York, 1962.

[32] E.N. Dancer, On the ranges of certain weakly nonlinear elliptic partial differential equations, J. Math. Pures Appl. 57 (1978), 351-366.

[33] C. De Coster, La méthode des sur et sous solutions dans l'étude de problèmes aux limites, Thèse de Doctorat, U.C.L., Louvain-la-Neuve (1994).

[34] C. De Coster, Pairs of positive solutions for the one-dimensional p-laplacian, Nonlinear Anal. T.M.A. 23 (1994), 669-681.

[35] C. De Coster, M.R. Grossinho and P. Habets, On pairs of positive solutions for a singular boundary value problem, to appear in Applicable Analysis.

[36] **C. De Coster** and **P. Habets**, *A two parameters Ambrosetti-Prodi problem*, to appear in Portugaliæ Mathematica.

[37] **D.G. de Figueiredo**, *Lectures on boundary value problems of the Ambrosetti-Prodi type*, Atas 12e Semin. Brasileiro Analise, Sao Paulo, (1980), 230-291.

[38] **D.G. de Figueiredo**, *Positive solutions of semilinear elliptic problems*, Lecture Notes in Math., vol. 957, Springer-Verlag, Berlin, 1982, 34-87.

[39] **D.G. de Figueiredo**, *Positive solutions for some classes of semilinear elliptic problems*, Proceedings of Symposia in Pure Mathematics **45** (1986), 371-379.

[40] **D.G. de Figueiredo** and **P.L. Lions**, *On pairs of positive solutions for a class of semilinear elliptic problems*, Indiana University Math. J. **34** (1985), 591-606.

[41] **D.G. de Figueiredo** and **W.N. Ni**, *Perturbations of second order linear elliptic problems by nonlinearities without Landesman-Lazer condition*, Nonlinear Anal., T.M.A. **5** (1979), 629-634.

[42] **D.G. de Figueiredo** and **S. Solimini**, *A variational approach to superlinear elliptic problems*, Comm. P.D.E. **9** (1984), 699-717.

[43] **J. Deuel** and **P. Hess**, *A criterion for the existence of solutions of non-linear elliptic boundary value problems*, Proc. Royal Soc. Edinburgh **74A** (1974/75), 49-54.

[44] **J.I. Diaz**, *Nonlinear differential equations and free boundaries. Vol 1. Elliptic equations*, Pitman Research Notes in Math. **106** (1985).

[45] **H. Epheser**, *Über die existenz der lösungen von randwertaufgaben mit gewöhnlichen, nichtlinearen differentialgleichungen zweiter ordnung*, Math. Zeitschr. **61** (1955), 435-454.

[46] **C. Fabry**, personal communication.

[47] **C. Fabry** and **P. Habets**, *The Picard boundary value problem for nonlinear second order vector differential equations*, J. Diff. Equ. **42** (1981), 186-198.

[48] **C. Fabry** and **P. Habets**, *Upper and lower solutions for second-order boundary value problems with nonlinear boundary conditions*, Nonlinear Anal. T.M.A. **10** (1986), 985-1007.

[49] **A. Fonda**, *On the existence of periodic solutions for scalar second order differential equations when only the asymptotic behaviour of the potential is known*, Proc. A.M.S **119** (1993), 439-445.

[50] **A. Fonda** and **J. Mawhin**, *Quadratic forms, weighted eigenfunctions and boundary value problems for nonlinear second order ordinary differential equations*, Proc. R. Soc. Edinburgh **112A** (1989), 145-153.

[51] **G. Fournier** and **J. Mawhin**, *On periodic solutions of forced pendulum-like equations*, J. Diff. Equ. **60** (1985), 381-395.

[52] **S. Fučík**, *Solvability of nonlinear equations and boundary value problems*, Reidel, Dordrecht, 1980.

[53] **J.A. Gatica, V. Oliker** and **P. Waltman**, *Singular nonlinear boundary value problems for second order ordinary differential equations*, J. Diff. Equ. **79** (1989), 62-78.

[54] **J.P. Gossez** and **P. Omari**, *Periodic solutions of a second order ordinary differential equation : a necessary and sufficient condition for nonresonance*, J. Diff. Equ. **94** (1991), 67-82.

[55] **J.P. Gossez** and **P. Omari**, *A necessary and sufficient condition of nonresonance for a semilinear Neumann problem*, Proc. A.M.S. **114** (1992), 433-442.

[56] **J.P. Gossez** and **P. Omari**, *Non-ordered lower and upper solutions in semilinear elliptic problems*, Comm. P. D. E. **19** (1994), 1163-1184.

[57] **J.P. Gossez** and **P. Omari**, *On a semilinear elliptic Neumann problem with asymmetric nonlinearities*, Trans. A.M.S. **347** (1995), 2553-2562.

[58] **V.V. Gudkov** and **A.J. Lepin**, *On necessary and sufficient conditions for the solvability of certain boundary-value problems for a second-order ordinary differential equation*, Dokl. Akad. Nauk SSSR **210** (1973), 800-803.

[59] **Z. Guo**, *Solvability of some singular nonlinear boundary value problems and existence of positive radial solutions of some nonlinear elliptic problems*, Nonlinear Anal. T.M.A. **16** (1991), 781-790.

[60] **P. Habets** and **M. Laloy**, *Perturbations singulières de problèmes aux limites: Les sur- et sous-solutions*, Séminaire de Mathématique Appliquée et Mécanique **76**, U.C.L. (1974).

[61] **P. Habets** and **P. Omari**, *Existence and localization of solutions of second order elliptic problems using lower and upper solutions in the reversed order*, to appear in Topological Methods in Nonlinear Analysis.

[62] **P. Habets** and **L. Sanchez**, *Periodic solutions of some Liénard equations with singularities*, Proc. A.M.S. **109** (1990), 1035-1044.

[63] P. Habets and F. Zanolin, *Upper and lower solutions for a generalized Emden-Fowler equation*, J. Math. Anal. Appl. **181** (1994), 684-700.

[64] P. Habets and F. Zanolin, *Positive solutions for a class of singular boundary value problem*, Boll. U.M.I. **(7) 9-A** (1995), 273-286.

[65] G. Harris, *A nonlinear Dirichlet problem with nonhomogeneous boundary data*, Applicable Anal. **33** (1989), 169-182.

[66] P. Hess, *An antimaximum principle for linear elliptic equations with an indefinite weight function*, J. Diff. Equ. **41** (1981), 369-374.

[67] S.I. Hudjaev, *Boundary problems for certain quasi-linear elliptic equations*, Soviet Math. Dokl. **5** (1964), 188-192.

[68] R. Iannacci, M.N. Nkashama and J.R. Ward, Jr., *Nonlinear second order elliptic partial differential equations at resonance*, Trans. A.M.S. **311** (1989), 711-726.

[69] F. Inkmann, *Existence and multiplicity theorems for semilinear elliptic equations with nonlinear boundary conditions*, Indiana Univ. Math. J. **31** (1982), 213-221.

[70] L.K. Jackson, *Subfunctions and second-order ordinary differential inequalities*, Advances in Math. **2** (1967), 307-363.

[71] J. Janus and J. Myjak, *A generalized Emden-Fowler equation with a negative exponent*, Nonlinear Analysis T.M.A. **23** (1994), 953-970.

[72] R. Kannan and K. Nagle, *Forced oscillations with rapidly vanishing nonlinearites*, Proc. A.M.S. **111** (1991), 385-393.

[73] R. Kannan and R. Ortega, *An asymptotic result in forced oscillations of pendulum-type equations*, Applicable Analysis **22** (1986), 45-53.

[74] J. Kazdan and F. Warner, *Remarks on some quasilinear elliptic equations*, Comm. Pure Appl. Math. **28** (1975), 567-597.

[75] H.B. Keller and D.S. Cohen, *Some positone problems suggested by non-linear heat generation*, J. Math. Mech. **16** (1967), 1361-1376.

[76] H.W. Knobloch, *Eine neue methode zur approximation periodischer lösungen nicht-linearer differentialgleichungen zweiter ordnung*, Math. Zeitschr. **82** (1963), 177-197.

[77] Y.S. Kolesov, *Periodic solutions of quasilinear parabolic equations of second order*, Trans. Moscow Math. Soc. **21** (1970), 114-146.

[78] **E.M. Landesman** and **A.C. Lazer**, *Nonlinear perturbations of linear elliptic boundary value problems at resonance*, J. Math. Mech. **19** (1970), 609-623.

[79] **A.C. Lazer** and **P.J. McKenna**, *On the number of solutions of a nonlinear Dirichlet problem*, J. Math. Anal. Appl. **84** (1981), 282-294.

[80] **A.C. Lazer** and **P.J. McKenna**, *On a conjecture related to the number of solutions of a nonlinear Dirichlet problem*, Proc. Royal Soc. Edinburgh **95A** (1983), 275-283.

[81] **A.C. Lazer** and **P.J. McKenna**, *On a conjecture on the number of solutions of a nonlinear Dirichlet problem with jumping nonlinearity*, in "Trends in theory of nonlinear differential equations (Arlington Texas 1982) L. N. Pure and Appl. Math. **90**, Dekker, New-York, (1984), 301-313.

[82] **A.C. Lazer** and **S. Solimini**, *On periodic solutions of nonlinear differential equations with singularities*, Proc. A.M.S. **99** (1987), 109-114.

[83] **S. Leela**, *Monotone method for second order periodic boundary value problems*, Nonlinear Analysis, T.M.A. **7** (1983), 349-355.

[84] **P.L. Lions**, *On the existence of positive solutions of semilinear elliptic equations*, SIAM Review **24** (1982), 441-467.

[85] **A.G. Lomtatidze**, *Positive solutions of boundary value problems for second order ordinary differential equations with singular points*, Differentsial'nye Uravneniya **23** (1987), 1685-1692.

[86] **J. Mawhin**, *Compacité, monotonie et convexité dans l'étude de problèmes aux limites semi-linéaires*, Séminaire d'analyse moderne 19, Université de Sherbrooke (1981).

[87] **J. Mawhin**, *Periodic oscillations of forced pendulum-like equations*, in "Ordinary and Partial Diff. Equ., Proceed. Dundee 1982", Lectures Notes in Math. 964, Springer, Berlin (ed: Everitt, Sleeman) (1982), 458-476.

[88] **J. Mawhin**, *Boundary value problems with nonlinearities having infinite jumps*, Comment. Math. Univ. Carolin. **25** (1984), 401-414.

[89] **J. Mawhin**, *Points fixes, points critiques et problèmes aux limites*, Sém. de Math. Supérieures, Univ. Montréal, (1985).

[90] **J. Mawhin**, *Problèmes de Dirichlet variationnels non linéaires*, Sém. de Math. Supérieures, Univ. Montréal, (1987).

[91] **J. Mawhin**, *Recent results on periodic solutions of the forced pendulum equation*, Rend. Ist. Matem. Univ. Trieste **19** (1987), 119-129.

[92] **J. Mawhin**, *The forced pendulum: A paradigm for nonlinear analysis and dynamical systems*, Expo. Math. **6** (1988), 271-287.

[93] **J. Mawhin**, *Boundary value problems for nonlinear ordinary differential equations: from successive approximations to topology*, in "Development of Mathematics, 1900-1950" (ed: J.P. Pier), Birkhauser, Basel (1994), 445-478.

[94] **J. Mawhin** and **J.R. Ward**, *Nonresonance and existence for nonlinear elliptic boundary value problems*, Nonlinear Anal. T.M.A. **5** (1981), 677-684.

[95] **J. Mawhin** and **J.R. Ward**, *Nonuniform nonresonance conditions at the two first eigenvalues for periodic solutions of forced Lienard and Duffing equations*, Rocky Mountain J. Math. **12** (1982), 643-654.

[96] **J. Mawhin** and **M. Willem**, *Multiple solutions of the periodic boundary value problem for some forced pendulum-type equations*, J. Diff. Equ. **52** (1984), 264-287.

[97] **E.J. McShane**, *Integration*, Princeton Univ. Press., Princeton (1944).

[98] **M. Nagumo**, *Über die differentialgleichung $y'' = f(t, y, y')$*, Proc. Phys-Math. Soc. Japan **19** (1937), 861-866.

[99] **M. Nagumo**, *On principally linear elliptic differential equations of the second order*, Osaka Math. J. **6** (1954), 207-229.

[100] **J.J. Nieto**, *Nonlinear second order boundary value problems with Carathéodory function*, Applicable Analysis **34** (1989), 111-128.

[101] **P. Omari**, *Non-ordered lower and upper solutions and solvability of the periodic problem for the Liénard and the Rayleigh equations*, Rend. Ist. Mat. Univ. Trieste **20** (1988), 54-64.

[102] **P. Omari** and **M. Trombetta**, *Remarks on the lower and upper solution method for second and third-order periodic boundary value problem*, Applied Math. Comput. **50** (1992), 71-82.

[103] **P. Omari** and **W. Ye**, *Necessary and sufficient conditions for the existence of periodic solutions of second order ODE with singular nonlinearities*, Diff. Int. Equ. **8** (1995), 1843-1858.

[104] **P. Omari** and **F. Zanolin**, *Infinitely many solutions of a quasilinear elliptic problem with an oscillatory potential*, preprint SISSA 1995.

[105] **L.C. Piccinini, G. Stampacchia** and **G. Vidossich**, *Ordinary differential equations in $\mathbb{R}^n$: Problems and methods*, Applied Math. Sciences 39, Springer-Verlag (1984).

[106] **N. Rouche** and **J. Mawhin**, *Equations différentielles ordinaires*, Masson, Paris, (1973).

[107] **D. Sattinger**, *Monotone methods in nonlinear elliptic and parabolic boundary value problems*, Indiana Univ. Math. J. **21** (1972), 979-1000.

[108] **D. Sattinger**, *Topics in stability and bifurcation theory*, Lecture Notes vol. 309, Springer, Berlin, 1973.

[109] **K. Schmitt**, *Bounded solutions of nonlinear second order differential equations*, Duke Math. J. **36** (1969), 237-244.

[110] **K. Schmitt**, *Boundary value problems for quasilinear second order elliptic equations*, Nonlinear Anal. T.M.A. **2** (1978), 263-309.

[111] **K. Schrader**, *Differential inequalities for second and third order equations*, J. Differential Equations **23** (1977), 203-215.

[112] **G. Scorza Dragoni**, *Il problema dei valori ai limiti studiato in grande per gli integrali di una equazione differenziale del secondo ordine*, Giornale di Mat (Battaglini) **69** (1931), 77-112.

[113] **G. Scorza Dragoni**, *Il problema dei valori ai limiti studiato in grande per le equazioni differenziali del secondo ordine*, Math. Ann. **105** (1931), 133-143.

[114] **G. Scorza Dragoni**, *Su un problema di valori ai limite per le equazioni differenziali ordinarie del secondo ordine*, Rend. Semin. Mat. R. Univ. Roma **2** (1938), 177-215. Aggiunta, ibid, 253-254.

[115] **G. Scorza Dragoni**, *Elementi uniti di transformazioni funzionali e problemi di valori ai limiti*, Rend. Semin. Mat. R. Univ. Roma **2** (1938), 255-275.

[116] **G. Scorza Dragoni**, *Intorno a un criterio di esistenza per un problema di valori ai limiti*, Rend. Semin. R. Accad. Naz. Lincei **28** (1938), 317-325.

[117] **L.F. Shampine** and **G.M. Wing**, *Existence and Uniqueness of Solutions of a Class of Nonlinear Elliptic Boundary Value Problems*, J. Math. and Mech. **19** (1970), 971-979.

[118] **R. Shivaji**, *A remark on the existence of three solutions via sub-super solutions*, Nonlinear Analysis and Applications (Ed: V. Lakshmikantham), Dekker Inc, New York and Basel, (1987), 561-566.

[119] **S.D. Taliaferro**, *A nonlinear singular boundary value problem*, Nonlinear Analysis T.M.A. **3** (1979), 897-904.

[120] **A. Tineo**, *Existence of two periodic solutions for the periodic equation* $\ddot{x} = g(t, x)$, J. Math. Anal. Appl. **156** (1991), 588-596.

[121] **A. Tineo**, *Existence theorems for a singular two-point Dirichlet problem*, Nonlinear Analysis T.M.A. **19** (1992), 323-333.

BOUNDARY VALUE PROBLEMS FOR QUASILINEAR SECOND ORDER DIFFERENTIAL EQUATIONS

R. Manásevich

University of Chile, Santiago, Chile

and

K. Schmitt

University of Utah, Salt Lake City, UT, USA

Abstract

This paper constitutes a survey of recent results on eigenvalue problems for nonlinear boundary value problems which arise in the study of radial solutions of quasilinear elliptic partial differential equations. Typical examples of such problems are boundary value problems for perturbations of the p-Laplacian.

1 Introduction

We are concerned with boundary value problems for nonlinear ordinary differential which arise in the search of radial solutions of nonlinear partial differential equations. The types of nonlinear partial differential equations we have in mind arise in a multitude of applied areas, such as the study of porous media, elasticity theory, plasma problems, astrophysics, to name only a few (see e.g. [26], [28], [39], [41], [48], [49], [57], and [65]).

The boundary value problems we are interested in belong to the following class of problems:

Let

$$A : \mathbb{R}^N \to \mathbb{R}^N, \ N \geq 1,$$

*AMS subject classification:*34A47, 34B15, 34C15, 35J25, 35J35, 35J70
Key words and phrases: radial solutions, quasilinear elliptic, p-Laplacian, bifurcation

be a continuous mapping that behaves asymptotically like

$$A(v) \cdot v \approx |v|^p,$$

where $p \in (1, \infty)$, more specifically,

$$A(v) \cdot v \geq \alpha |v|^p,$$

and

$$|A(v)| \leq \beta |v|^{p-1},$$

where α and β are positive constants. Also A is assumed to be strictly monotone with respect to v i.e.

$$(A(v_1) - A(v_2)) \cdot (v_1 - v_2) > 0, \ v_1 \neq v_2.$$

We consider the boundary value problem

$$\begin{aligned} -\mathrm{div} A(\nabla u) &= f(\lambda, u), \ x \in \Omega, \\ u &= 0, \ x \in \partial\Omega, \end{aligned} \tag{1}$$

where Ω is a bounded domain in $\mathbb{R}^N$.

An important special case of the above is when the operator A has the form

$$A(\nabla u) = a(|\nabla u|)\nabla u$$

and when Ω is either a ball or a spherical shell (an annular domain) in $\mathbb{R}^N$ (see the above quoted references). In these cases it is of interest to seek solutions which only depend on the radial variable, i.e. solutions u such that $u(x) = u(|x|) = u(r)$. Such solutions are then solutions of the following boundary value problems for ordinary differential equations

$$\begin{aligned} \left[r^{N-1}\phi(u')\right]' + r^{N-1}g(\lambda, u) &= 0, \ r \in (0, R), \\ u'(0) = 0, \ u(R) &= 0, \end{aligned} \tag{2}$$

in case Ω is a ball of radius R in $\mathbb{R}^N$ and

$$\begin{aligned} \left[r^{N-1}\phi(u')\right]' + r^{N-1}g(\lambda, u) &= 0, \ r \in (a, b), \\ u(a) = 0, \ u(b) &= 0, \end{aligned} \tag{3}$$

in case $\Omega = \{x \in \mathbb{R}^N : a < |x| < b\}$, where ϕ is an increasing homeomorphism of $\mathbb{R}$, with $\phi(0) = 0$.

It is these classes of boundary value problems which we shall study here for various classes of nonlinearities ϕ and g (it usually will be assumed that $ug(\lambda, u) \geq 0$). We note that in either case the differential equation may also be written as

$$[\phi(u')]' + \frac{N-1}{r}\phi(u) + g(\lambda, u) = 0, \tag{4}$$

which, in some instances, will be a more convenient form.

This paper, being mostly a survey, reports about several recent papers, where tools from nonlinear analysis have been used to analyze boundary value problems of the types (2) and (3) and we shall rely on the papers [17], [18], [30], [32] and [31] for our main sources.

The tools employed are the tools of nonlinear functional analysis, specifically we employ results from various sources, such as [1], [8], [9], [10], [36], [41], [42], [43], [50], [59], [62], [64].

The paper is organized as follows. The first part is concerned with the problem (2). Here we first consider the case of nonlinearities g which grow at the same rate as the nonlinear term ϕ and are linear with respect to the parameter λ, i.e. a case which has as its analogue Sturm-Liouville type eigenvalue problems for second order linear ordinary differential equations. We then continue with problems where g grows superlinearly (but subcritically) with respect to ϕ. In the second part we consider problem (3), i.e. boundary value problems on an annular domain, where the term g grows superlinearly with respect to ϕ. The case where g grows linearly with respect to ϕ need not be considered, as its treatment is similar (somewhat easier) as in the case of the ball. The last section is devoted to the true partial differential equations case, i.e., we consider eigenvalue problems for quasilinear equations of the above types on general domains.

For sources containing similar problems and results we refer the reader to the following: [3], [15], [20], [34], [40], [48], [51], and [53].

2 Boundary value problems on a ball

2.1 Equivalent integral equation

In this section we shall derive an integral equation whose solution set is the set of positive solutions of (2). A quick calculation shows that finding positive solutions to problem (2) is equivalent to finding nontrivial solutions to the problem

$$\left[r^{N-1}\phi(u')\right]' + r^{N-1}g(\lambda, |u|) = 0, \ r \in (0, R),$$
$$u'(0) = 0, \ u(R) = 0. \tag{5}$$

Let $C_\#$ denote the closed subspace of $C[0, R]$ defined by

$$C_\# = \{u \in C[0, R] |\ u(R) = 0\}.$$

Then $C_\#$ is a Banach space for the norm $\| \ \| := \| \ \|_\infty$.

Let $u(r)$ be a solution of (5). By integrating the equation in (5) we see that that $u(r)$ satisfies the integral equation

$$u(r) = \int_r^R \phi^{-1}\left[\frac{1}{s^{N-1}}\int_0^s \xi^{N-1}g(\lambda, |u(\xi)|)d\xi\right]ds. \tag{6}$$

Let us define $T(\cdot, \cdot) : C_\# \times [0, \infty) \to C_\#$ by

$$T(u, \lambda)(r) = \int_r^R \phi^{-1} \left[\frac{1}{s^{N-1}} \int_0^s \xi^{N-1} g(\lambda, |u(\xi)|) d\xi \right] ds. \tag{7}$$

Clearly T is well defined and fixed points of $T(\cdot, \lambda)$ will provide solutions of (5).

2.2   Eigenvalue problems

In this section we shall consider the problem

$$\begin{aligned} &\left[r^{N-1} \phi(u') \right]' + \lambda r^{N-1} \psi(u) = 0, \ r \in (0, R), \\ &u'(0) = 0, \ u(R) = 0, \end{aligned} \tag{8}$$

where ϕ is an increasing homeomorphism of $\mathbb{R}$ and ψ is a nondecreasing function with $\phi(0) = \psi(0) = 0$ and for any $\sigma > 0$

$$A(\sigma) \leq \frac{\phi(\sigma|x|)}{|\psi(x)|} \leq B(\sigma), \tag{9}$$

where $A(\sigma)$ and $B(\sigma)$ are positive constants depending on σ only. We seek the existence of values of λ such that the problem has a positive solution, such values of λ will be called principal eigenvalues, and we seek the existence of values of λ such that the problem has a nontrivial signchanging solution, such values of λ will be called higher eigenvalues.

The problem is a natural extension of the eigenvalue problem for the p-Laplacian $(\phi(u) = \psi(u) = |u|^{p-2}u, \ p > 1)$ considered in [3], [20], [15] among others. Our results presented here are based on the work in [31], [32].

2.2.1   On principal eigenvalues

Let us denote by

$$S = \{(\lambda, u) \in \mathbb{R}_+ \times C_\# : (\lambda, u) \text{ solves } (8), \ u(r) > 0, \ r \in (0, R)\}. \tag{10}$$

We establish the following theorem.

Theorem 2.1 *Let S be defined by (10) and let ϕ and ψ be homeomorphisms satisfying (9). Then $S \neq \emptyset$ and there exist numbers $\lambda_1 > 0$ and $\lambda_2 > 0$ such that $(\lambda, u) \in S$ implies $\lambda_1 \leq \lambda \leq \lambda_2$. Further more there exists a continuum $\mathcal{C} \subset \bar{S}$ which is unbounded in $[\lambda_1, \lambda_2] \times C_\#$ and bifurcates from $[\lambda_1, \lambda_2] \times \{0\}$.*

Proof We consider the equivalent operator equation. The operator T in this case has the form

$$T(u, \lambda)(r) = \int_r^R \phi^{-1} \left[\frac{1}{s^{N-1}} \int_0^s \xi^{N-1} \lambda \psi(|u(\xi)|) d\xi \right] ds. \tag{11}$$

We shall establish that

$$\deg_{LS}(I - T(\cdot, 0), B(0, R_1), 0) = 1 \tag{12}$$

and

$$\deg_{LS}(I - T(\cdot, \lambda_2), B(0, R_1), 0) = 0, \tag{13}$$

for some $\lambda_2 > 0$ and any $R_1 > 0$, where $\deg_{LS}$ denotes Leray-Schauder degree. That (12) holds follows, since $T(\cdot, 0) = 0$. To show that (13) holds, we consider the operator $T_\epsilon : C_\# \times [0, \infty) \to C_\#$, by

$$T_\epsilon(u, \lambda)(r) = \int_r^R \phi^{-1}\left[\frac{1}{s^{N-1}} \int_0^s \xi^{N-1} \lambda(\phi(|u(\xi)|) + \epsilon) d\xi\right] ds, \tag{14}$$

where $\epsilon > 0$ is a constant. We have that T_ϵ sends bounded sets of $C_\# \times [0, \infty)$ into bounded sets of $C_\#$. Moreover, T_ϵ is a completely continuous operator and $T_\epsilon(\cdot, 0) = 0$. Furthermore

$$\deg_{LS}(I - T(\cdot, \lambda), B(0, R_1), 0) = \deg_{LS}(I - T_\epsilon(\cdot, \lambda), B(0, R_1), 0), \tag{15}$$

for all small ϵ, whenever $\deg_{LS}(I - T(\cdot, \lambda), B(0, R_1), 0)$ is defined. Thus (13) will be true if we can show that there exists $\bar{\lambda}$ such that (u, λ) a solution of

$$u = T_\epsilon(\lambda, u)$$

implies $\lambda \leq \bar{\lambda}$ and that this number is independent of ϵ for $0 < \epsilon \leq \epsilon_0$. Thus, assume that u satisfies

$$-r^{N-1}\phi(u'(r)) = \int_0^r \xi^{N-1} \lambda(\psi(|u(\xi)|) + \epsilon) d\xi \geq 0$$

and

$$u(r) = \int_r^R \phi^{-1}\left[\frac{1}{s^{N-1}} \int_0^s \xi^{N-1} \lambda(\psi(|u(\xi)|) + \epsilon) d\xi\right] ds \geq 0.$$

Hence, $u'(r) \leq 0$ and $u(r) \geq 0$ for all $r \in [0, R]$. Also for $r \in \left[\frac{R}{4}, \frac{3R}{4}\right]$

$$u(r) \geq \int_r^R \phi^{-1}\left[\frac{1}{s^{N-1}} \int_0^s \xi^{N-1} \lambda\psi(|u(\xi)|) d\xi\right] ds.$$

Thus, for all $r \in \left[\frac{R}{4}, \frac{3R}{4}\right]$, we have that

$$u(r) \geq \frac{R}{4}\phi^{-1}\left[\frac{\lambda R}{N 4 3^{N-1}}\psi(u(r))\right]$$

or equivalently,

$$\frac{\phi\left(\frac{4}{R}u(r)\right)}{\psi(u(r))} \geq \frac{\lambda R}{N 4 3^{N-1}}. \tag{16}$$

It is clear that (16) cannot hold for λ large, as follows from condition (9). We have thus, using the homotopy invariance of the Leray-Schauder degree established that $S \neq \emptyset$ and the existence of λ_2, as asserted. The existence of an unbounded continuum, as claimed, follows from the global Krasnosel'skii-Rabinowitz bifurcation theorem (see [19]). To finish the proof, we need to establish the existence of λ_1. Thus let u be a solution with $u(0) = d$. It follows that

$$d \leq \int_0^R \phi^{-1} \left[\int_0^R \lambda \psi(d) \frac{s}{N} \right] ds, \tag{17}$$

which implies

$$\frac{\phi(d/R)}{\psi(d)} \leq \lambda \frac{R}{N}.$$

We now use (9) to obtain a lower bound on λ.

The above calculations also imply the following corollary:

Corollary 2.1 *Let (u, λ) be a solution of (8) with $u(0) = d$ and let $\theta \in (0, 1)$ be fixed. Let $r_0 \in (0, R)$ be such that $u(r_0) = \theta d$. Then*

$$r_0 \geq \frac{N}{\lambda} A \left(\frac{1 - \theta}{R} \right). \tag{18}$$

Proof Using earlier calculations we obtain the following

$$\theta d = \int_{r_0}^R \phi^{-1} \left[\frac{1}{s^{N-1}} \int_0^s \xi^{N-1} \lambda \psi(|u(\xi)|) d\xi \right] ds, \tag{19}$$

and hence

$$(1 - \theta)d = \int_0^{r_0} \phi^{-1} \left[\frac{1}{s^{N-1}} \int_0^s \xi^{N-1} \lambda \psi(|u(\xi)|) d\xi \right] ds,$$

from which follows

$$(1 - \theta)d \leq \int_0^{r_0} \phi^{-1} \left[\frac{1}{s^{N-1}} \int_0^s \xi^{N-1} \lambda \psi(d) d\xi \right] ds,$$

and

$$(1 - \theta)d \leq R \phi^{-1} \left[\frac{\lambda \psi(d) r_0}{N} \right],$$

which implies the conclusion.

The result just proved has the following consequence.

Corollary 2.2 *Let $\{(u_n, \lambda_n)\}$ be a sequence of solutions of (8) with $u_n(0) = d_n$. Then $u_n(r) \to \infty$ uniformly with respect to r in compact subintervals of $[0, R)$.*

Proof Since u_n is given by

$$u_n(r) = \int_r^R \phi^{-1}\left[\frac{1}{s^{N-1}}\int_0^s \xi^{N-1}\lambda_n\psi(|u_n(\xi)|)d\xi\right]ds,$$

we obtain for $r \geq r_0$ (viz. Corollary 2.1) that

$$u_n(r) \geq \int_r^R \phi^{-1}\left[\frac{1}{s^{N-1}}\int_0^{r_0} \xi^{N-1}\lambda_n\psi(\theta d_n)d\xi\right]ds.$$

The conclusion follows from this inequality.

2.2.2 Monotonicity of principal eigenvalues

In this section we shall show that

$$\lambda_1^-(R) = \inf\{\lambda : \lambda \text{ principal eigenvalue of (8)}\}$$

is a nondecreasing function of R. To this end we shall need the following elementary properties of the operator T.

We note that the space $C_\#$ is a partially ordered Banach space with respect to the partial order $\leq$, i.e. for $u, v \in C_\#$, $u \leq v$ whenever $u(r) \leq v(r)$, $r \in [0, R]$. Further, if $[u, v] = \{w \in C_\# : u \leq w \leq v\}$ is an order interval in $C_\#$, then it is a bounded closed set in $C_\#$.

We shall also use the notation

$$\lambda_1^+(R) = \sup\{\lambda : \lambda \text{ principal eigenvalue of (8)}\}$$

Proposition 2.1 *The operator T is monotone with respect to the above partial order in $C[0, R]$ and hence in $C_\#$ and also monotone with respect to λ.*

From this proposition and the complete continuity of T immediately follows the following fixed point result.

Proposition 2.2 *Assume there exists $[\alpha, \beta] \subset C[0, R]$ such that*

$$T(\lambda, \cdot) : [\alpha, \beta] \to [\alpha, \beta].$$

Then $T(\lambda, \cdot)$ has a fixed point $u \in C_\# \cap [\alpha, \beta]$.

We note that the hypotheses of Proposition 2.2 will hold, whenever we can find a pair $\{\alpha, \beta\} \subset C[0, R]$ such that

$$\alpha \leq \beta$$

and

$$\alpha \leq T(\lambda, \alpha), \ T(\lambda, \beta) \leq \beta.$$

Using these facts we can now establish the following result.

Theorem 2.2 λ_1^- *is a nonincreasing function of R.*

Proof Assume there exist R_1, R_2, $R_1 < R_2$ such that $\lambda_1^-(R_1) < \lambda_1^-(R_2)$. Then there exists μ, $\lambda_1^-(R_1) \leq \mu < \lambda_1^-(R_2)$, such that (8) has a nontrivial solution $\tilde{\alpha}$ for $R = R_1$ and $\lambda = \mu$. Furthermore, there exists $\nu \geq \lambda_1^-(R_2)$ and a nontrivial solution β of (2.1) for $R = R_2$ and $\lambda = \nu$, with $\beta(0) = d$ as large as desired. It follows from Corollary 2.2 that for d sufficiently large $\beta(r) > \tilde{\alpha}(r)$, $0 \leq r \leq R_1$. Define

$$\alpha = \begin{cases} \tilde{\alpha}, & 0 \leq r \leq R_1 \\ 0, & R_1 \leq r \leq R_2. \end{cases}$$

Then the operator $T(\mu, \cdot)$ for $R = R_2$ satisfies in the space $C[0, R_2]$

$$\alpha \leq T(\mu, \alpha), \ T(\mu, \beta) \leq \beta,$$

as may easily be verified. Thus by Proposition 2.2 this operator will have a fixed point in $[\alpha, \beta]$, contradicting that $\mu < \lambda_1^-(R_2)$.

Remark 2.1 *We have stablished above that for every level d (8) has a principal eiegnevalue $\lambda(d)$ with an associated eigenfucntion u such that $u(0) = d$. We hence may define*

$$\lambda_1^-(d, R) = \inf\{\lambda(d) : \lambda(d) \ \text{principal eigenvalue of (8) on level d}\}$$

and

$$\gamma_1(R) = \liminf_{d \to 0} \lambda_1^-(d, R)$$

and

$$\Gamma_1(R) = \liminf_{d \to \infty} \lambda_1^-(d, R).$$

It follows by an argument similar to the one used above in the proof of Theorem 2.2 that both $\gamma_1(R)$ and $\Gamma_1(R)$ are nondecreasing functions of R also.

Remark 2.2   *1. Theorem 2.2 implies that problem (8) has no nontrivial solutions for $\lambda < \lambda_1^-(R)$.*

2. Solutions of (8) are a priori bounded for λ in compact subintervals of $(-\infty, \lambda_1^-(R))$.

3. Our calculations and results also imply that for $\delta > 0$, small, the degrees

$$\deg_{LS}(I - T(\cdot, a), B(0, \delta), 0)$$

$$\deg_{LS}(I - T(\cdot, b), B(0, \delta), 0)$$

are defined for $a < \lambda_1^-(R)$ and $b > \lambda_1^+(R)$ and equal 1 and 0, respectively. Hence it follows from global bifurcation theory (see [19]) that an unbounded continuum of positive solutions of (23) will bifurcate from $[\lambda_1^-(R), \lambda_1^+(R)]$.

2.2.3 On the principal eigenvalue of the p-Laplacian

The above considerations allow for an easy and straightforward proof of the fact that for the p-Laplacian (see for example, [3], [4], [5], and [20]) the eigenvalue problem $(\phi_p(s) = |s|^{p-2}s)$

$$-\left[r^{N-1}\phi_p(u')\right]' = r^{N-1}\lambda\phi_p(u), \text{ in } (0, R)$$
$$u'(0) = 0 = u(R),$$

(20)

has a unique principal eigenvalue with all eigenfunctions being a constant multiple of a given one. To see this we observe that, because of the homogeneity of the problem, constant multiples of eigenfunctions are eigenfunctions. Hence, if $\lambda_1 < \lambda_2$ are principal eigenvalues of (20) with associated eigenfunctions u_1, u_2. It follows from the equation that $u_1'(R) \neq 0 \neq u_2'(R)$.

We next find constants $\alpha > 0$ and $\beta > 0$ such that

$$\alpha u_1(t) \leq u_2(t) \leq \beta u_1(t), \ 0 \leq t \leq R,$$

further, α may be chosen maximal and β minimal. This, on the other hand will imply that $\alpha = \beta$ and hence $u_2 = \alpha u_1$, which in turn implies that $\lambda_1 = \lambda_2$. (to make these arguments precise we employ arguments like the ones used later in section 3.

2.2.4 On higher eigenvalues

Let us next assume that $\psi : \mathbb{R} \to \mathbb{R}$ is also an odd increasing homeomorphism of $\mathbb{R}$, with $\psi(0) = 0$. Furthermore we will require that ϕ, ψ satisfy the asymptotic homogeneity conditions:

$$\lim_{s \to 0} \frac{\phi(\sigma s)}{\psi(s)} = \sigma^{p-1}, \quad \text{for all} \quad \sigma \in \mathbb{R}_+, \quad \text{for some } p > 1,$$

(21)

and

$$\lim_{s \to \pm\infty} \frac{\phi(\sigma s)}{\psi(s)} = \sigma^{q-1}, \quad \text{for all} \quad \sigma \in \mathbb{R}_+, \quad \text{for some } q > 1.$$

(22)

We note that if the pair ϕ, ψ satisfies the asymptotic homogeneity conditions (21) and (22), then the function ϕ satisfies both of these conditions with ψ replaced by ϕ and also ψ satisfies both of these conditions with ϕ replaced by ψ. Such conditions appear in the study of Dirichlet boundary value problems using blow-up techniques, (see [33]), in the one dimensional case using time mapping techniques, see [35], and in [29] where a very special case of the above class of functions was considered to obtain a local bifurcation result for positive solutions. We here recall the well-known fact that the eigenvalue problem (20) has a sequence $\{\lambda_n(p)\}_n$ of positive eigenvalues such that $\lambda_n(p) \to \infty$ as $n \to \infty$, and associated with each $\lambda_n(p)$ there is a one dimensional space spanned by a solution of (20) with exactly $n - 1$ simple zeros in $(0, R)$, (see [3], [21]).

2.2.5   Index calculations

Let us consider again the problem

$$-\left[r^{N-1}\phi(u')\right]' = r^{N-1}\lambda\psi(u), \text{ in } (0,R)$$
$$u'(0) = 0 = u(R). \tag{23}$$

A straightforward calculation shows, as was already done, that u is a solution to (23) if and only if u is a fixed point of the completely continuous operator $T^\lambda_{\phi\psi} : C[0,R] \to C[0,R]$ defined by

$$T^\lambda_{\phi\psi}(u)(r) = \int_r^R \phi^{-1}\left[\frac{1}{s^{N-1}}\int_0^s \xi^{N-1}\lambda\psi(u(\xi))d\xi\right]ds. \tag{24}$$

The main purpose of this section is to prove the following:

Lemma 2.1 *Consider the problem (23) and assume that ϕ,ψ are odd increasing homeomorphisms of $\mathbb{R}$, with $\phi(0) = 0, \psi(0) = 0$.*

- *If*

$$\lim_{s\to 0} \frac{\phi(\sigma s)}{\psi(s)} = \sigma^{p-1}, \quad \text{for all} \quad \sigma \in \mathbb{R}_+, \tag{25}$$

  then the Leray-Schauder degree of $I - T^\lambda_{\phi\psi}$ is defined for $B(0,\varepsilon)$, for all sufficiently small ε. Furthermore we have

$$\deg_{LS}(I - T^\lambda_{\phi\psi}, B(0,\varepsilon), 0) = \begin{cases} 1, & \text{if } \lambda < \lambda_1(p), \\ (-1)^m, & \text{if } \lambda \in (\lambda_m(p), \lambda_{m+1}(p)). \end{cases} \tag{26}$$

- *If*

$$\lim_{|s|\to\infty} \frac{\phi(\sigma s)}{\psi(s)} = \sigma^{q-1}, \quad \text{for all} \quad \sigma \in \mathbb{R}_+, \tag{27}$$

  then the Leray-Schauder degree for $I - T^\lambda_{\phi\psi}$ is defined for $B(0,M)$, for all sufficiently large M, and a similar formula to (26) holds, namely

$$\deg_{LS}(I - T^\lambda_{\phi\psi}, B(0,R), 0) = \begin{cases} 1, & \text{if } \lambda < \lambda_1(q) \\ (-1)^l, & \text{if } \lambda \in (\lambda_l(q), \lambda_{l+1}(q)), \end{cases} \tag{28}$$

  where $\{\lambda_l(q), l = 1, 2, \cdots\}$ is the set of eigenvalues of (20), with p replaced by q.

Proof We only give the proof for the case that (25) holds. Consider the completely continuous operator T^λ_p corresponding to the case $\phi = \psi = \phi_p$ given by

$$T_p^\lambda(u)(r) = \int_r^R \phi_p^{-1}\left[\frac{1}{s^{N-1}}\int_0^s \xi^{N-1}\lambda\phi_p(u(\xi))d\xi\right]ds. \qquad (29)$$

It is known, (see [21]), that for this operator (26) holds. Define the operator T^λ : $C[0,R] \times [0,1] \to \mathbb{R}_+$, by

$$T^\lambda(u,\tau) = \tau T_{\phi\psi}^\lambda(u) + (1-\tau)T_p^\lambda(u). \qquad (30)$$

We claim that the Leray Schauder degree for $I - T^\lambda(\cdot,\tau)$ is defined for $B(0,\varepsilon)$, in $C[0,R]$, for all small ε. Indeed, suppose this is not the case, then there exist sequences $\{u_n\}$, $\{\tau_n\}$ and $\{\varepsilon_n\}$ with $\varepsilon_n \to 0$, and $||u_n|| = \varepsilon_n$ such that

$$u_n = T^\lambda(u_n,\tau_n),$$

i.e. such that

$$u_n(r) = \tau_n \int_r^R \phi^{-1}\left[\frac{1}{s^{N-1}}\int_0^s \xi^{N-1}\lambda\psi(u_n(\xi))d\xi\right]ds$$
$$+(1-\tau_n)\int_r^R \phi_p^{-1}\left[\frac{1}{s^{N-1}}\int_0^s \xi^{N-1}\lambda\phi_p(u_n(\xi))d\xi\right]ds. \qquad (31)$$

Setting $\hat{u}_n = \frac{u_n}{\varepsilon_n}$ we have that $||\hat{u}_n|| = 1$ and

$$\hat{u}_n(r) = \left(\frac{\tau_n}{\varepsilon_n}\int_r^R \phi^{-1}\left[\frac{1}{s^{N-1}}\int_0^s \xi^{N-1}\lambda\psi(u_n(\xi))d\xi\right]ds\right)$$
$$+(1-\tau_n)\int_r^R \phi_p^{-1}\left[\frac{1}{s^{N-1}}\int_0^s \xi^{N-1}\lambda\phi_p(\hat{u}_n(\xi))d\xi\right]ds. \qquad (32)$$

We hence obtain that

$$\hat{u}_n'(r) = \left(\frac{\tau_n}{\varepsilon_n}\phi^{-1}\left[\frac{1}{r^{N-1}}\int_0^r \xi^{N-1}\lambda\psi(u_n(\xi))d\xi\right]\right)$$
$$+(1-\tau_n)\phi_p^{-1}\left[\frac{1}{r^{N-1}}\int_0^r \xi^{N-1}\lambda\phi_p(\hat{u}_n(\xi))d\xi\right] \qquad (33)$$

and conclude that the sequence $\{\hat{u}_n'\}$ is uniformly bounded. Thus $\{\hat{u}_n\}$ contains a uniformly convergent subsequence which we relabel as the original sequence. Let $\lim \hat{u}_n = \hat{u}$.

For each $n \in \mathbb{N}$ let us set

$$v_n(r) := \int_r^R h_n(s)ds,$$

where

$$h_n(s) = \frac{1}{\varepsilon_n}\Big(\phi^{-1}\Big[\frac{\phi(\varepsilon_n)}{s^{N-1}}\int_0^s \frac{\xi^{N-1}\lambda\psi(u_n(\xi))d\xi}{\phi(\varepsilon_n)}\Big]\Big).$$

By an application of the Lebesgue dominated convergence theorem and the fact that ϕ is asymptotically homogeneous of degree $p-1$ at zero, we obtain that $h_n \to h$ in $L^1(0,R)$, where

$$h(s) = \phi_p^{-1}\Big[\frac{\lambda}{s^{N-1}}\int_0^s \xi^{N-1}\phi_p(\hat{u}(\xi))d\xi\Big].$$

Set

$$v(r) = \int_r^R h(s)ds = \int_r^R \phi_p^{-1}\Big(\frac{\lambda}{s^{N-1}}\int_0^s \xi^{N-1}\phi_p(\hat{u}(\xi))d\xi\Big)ds.$$

Then, since

$$\|v_n - v\| \le \|h_n - h\|_{L^1(0,R)},$$

we conclude that $v_n \to v$ in $C[0,R]$.

Thus letting $n \to \infty$ in (32) and since we can assume, without loss of generality, that $\tau_n \to \tau_0$ as $n \to \infty$, we find that $\hat{u}$ satisfies

$$\hat{u} = T_p^\lambda(\hat{u}).$$

Since $\lambda \notin \{\lambda_m(p)\}$, it must be the case that $\hat{u} = 0$ (see [21]), yielding a contradiction.

Thus, by the properties of the Leray Schauder degree, we have that

$$\deg_{LS}(I - T_{\phi\psi}^\lambda, B(0,\varepsilon), 0) = \deg_{LS}(I - T_p^\lambda, B(0,\varepsilon), 0)), \tag{34}$$

for all ε sufficiently small, and hence (26) follows from Proposition 4.2 in [21].

We complete this section by applying Lemma 2.1 to obtain an existence result for nontrivial solutions.

Theorem 2.3 *Consider problem (23) and suppose that ϕ, ψ are odd increasing homeomorphisms of $\mathbb{R}$ with $\phi(0) = 0 = \psi(0)$, which satisfy (25) and (27) with $p \ne q$. Assume that for some $j \in \mathbb{N}$, $\lambda_j(p) \ne \lambda_j(q)$ and that $\lambda \in (A, B)$, where $A = \min\{\lambda_j(p), \lambda_j(q)\}$ and $B = \max\{\lambda_j(p), \lambda_j(q)\}$. Assume furthermore that (A, B) does not contain any other eigenvalues from $\{\lambda_n(p)\}$ or $\{\lambda_n(q)\}$. Then problem (23) has a nontrivial solution.*

Proof Assuming for example that $\lambda_j(q) < \lambda_j(p)$, it follows from Lemma 2.1 that

$$\deg_{LS}(I - T^\lambda_{\phi\psi}, B(0, \varepsilon), 0) = (-1)^j, \tag{35}$$

for $\varepsilon > 0$ small, and that

$$\deg_{LS}(I - T^\lambda_{\phi\psi}, B(0, M), 0) = (-1)^{j-1}, \tag{36}$$

for M large. Thus combining (35) and (36) with the excision property of the Leray-Schauder degree, we obtain that

$$\deg_{LS}(I - T^\lambda_{\phi\psi}, B(0, M) \setminus \overline{B(0, \varepsilon)}, 0) \neq 0, \tag{37}$$

yielding the existence of a nontrivial solution.

This theorem suggests -and one may prove this- the existence of a continuum of solutions to (23) connecting $(0, \lambda_j(p))$ with $(\infty, \lambda_j(q))$, for each $j \in \mathbb{N}$, generalizing the well known property for the homogeneous case.

3   On initial value problems

In this section we will prove some results for the initial value problem associated with perturbations of the problem (23) i.e.,

$$-\left[r^{N-1}\phi(u')\right]' = r^{N-1}\lambda\psi(u) + r^{N-1}f(r, u, \lambda), \text{ in}(0, R)$$
$$u(0) = d, \quad u'(0) = 0, \tag{38}$$

which will be needed later. Thoroughout we shall assume that $uf(r, u, \lambda) \geq 0$.

Proposition 3.1 *Suppose that $f(r, u, \lambda) = O(|\psi(u)|)$ near zero, uniformly for r and λ in bounded intervals. Then the only solution to the problem*

$$-\left[r^{N-1}\phi(u')\right]' = r^{N-1}\lambda\psi(u) + r^{N-1}f(r, u, \lambda), \text{ in } (0, R)$$
$$u(r_0) = 0, \quad u'(r_0) = 0,$$

with $r_0 \geq 0$ is the trivial one.

Proof Suppose u is a solution such that $u \not\equiv 0$ in the interval $[r_0, r_0 + \delta)$, for some $\delta > 0$. Then integrating the equation from r_0 to $r \in (r_0, r_0 + \delta]$, we find

$$-u(r) = \int_{r_0}^r \phi^{-1}\left(\frac{1}{s^{N-1}} \int_{r_0}^s \xi^{N-1}(\lambda\psi(u(\xi)) + f(\xi, u(\xi), \lambda)d\xi\right)ds.$$

Because of the assumptions on f there exists $\varepsilon > 0$ such that for δ small

$$|u(r)| \leq \int_{r_0}^r \phi^{-1}\left(\frac{1}{s^{N-1}} \int_{r_0}^s \xi^{N-1}(\lambda\psi(|u|_\delta) + \varepsilon\psi(|u|_\delta))d\xi\right)ds,$$

where $|u|_\delta$ denotes the sup norm of u in $[r_0, r_0 + \delta]$. Hence

$$|u|_\delta \leq \delta\phi^{-1}[(\lambda + \varepsilon)\psi(|u|_\delta|)\delta],$$

and

$$\frac{\phi(|u|_\delta/\delta)}{\phi(|u|_\delta)} \leq C\delta\frac{\psi(|u|_\delta)}{\phi(|u|_\delta)}, \tag{39}$$

where C is a constant independent of δ. For δ small we obtain a contradiction, since the left hand side of the inequality (39) exceeds 1 (recall the conditions (25), (27).) A similar argument applies for an interval of the form $[r_0 - \delta, r_0]$, in case $r_0 > 0$.

The next results extends to our situation a result of [31] (see also [52]). We refer to [32] for proofs.

Proposition 3.2 *Suppose that $f(r, u, \lambda) = O(|\psi(u)|)$ near zero, uniformly for r and λ in bounded intervals, then nontrivial solutions of the initial value problem (38) are oscillatory.*

Proposition 3.3 *Suppose that $f(r, u, \lambda) = O(|\psi(u)|)$ near zero, uniformly for r and λ in bounded intervals. Let ρ_d denote the first zero of a solution u of (38) such that $u(0) = d \neq 0$, then for all $K > 0$ there exists $\Lambda(K)$ such that for all $\lambda > \Lambda(K)$ and all $d \neq 0$ we have that $\rho_d \leq K$.*

Proposition 3.4 *Suppose that $f(r, u, \lambda) = O(|\psi(u)|)$ near zero, uniformly for r and λ in bounded intervals, then for all $L > 0$ and $j \in \mathbb{N}$ there exists $\Lambda_j(L) > 0$ such that for all $\lambda > \Lambda_j(L)$ and all $d \neq 0$ we have that $\rho_{d_j} \leq L$.*

4   Bifurcation of solutions

We begin this section by recalling what we mean by bifurcation at zero and at infinity (see [42], [59], and [61]). Let X be a Banach space with norm $\|\cdot\|$, and let $\mathcal{F} : X \times I \to \mathbb{R}$ be a completely continuous operator, where I is some real interval. Consider the equation

$$x = \mathcal{F}(x, \lambda). \tag{40}$$

Definition 1 Suppose that $\mathcal{F}(0, \lambda) = 0$ for all λ in I, and that $\hat{\lambda} \in I$. We say that $(0, \hat{\lambda})$ is a bifurcation point of (40) at zero if in any neighborhood of $(0, \hat{\lambda})$ in $X \times I$ there is a nontrivial solution of (40). Or equivalently, if there exist sequences $\{x_n \neq 0\}$ and $\{\lambda_n\}$ with $(x_n, \lambda_n) \to (0, \hat{\lambda})$ and such that (x_n, λ_n) satisfies (40) for each $n \in \mathbb{N}$.

Definition 2 We say that $(\infty, \hat{\lambda})$ is a bifurcation point of (40) at infinity, provided in any neighborhood of $(\infty, \hat{\lambda})$ in $X \times I$ there is a solution of (40). Equivalently if there exist sequences $\{x_n\}$ and $\{\lambda_n\}$ with $(\|x_n\|, \lambda_n) \to (\infty, \hat{\lambda})$ and such that (x_n, λ_n) satisfies (40) for each $n \in \mathbb{N}$.

Let u be a solution of the problem

$$-\left[r^{N-1}\phi(u')\right]' = r^{N-1}\lambda\psi(u) + r^{N-1}f(r, u, \lambda), \quad \text{in}(0, R) \tag{41}$$
$$u(R) = 0, \quad u'(0) = 0,$$

then u satisfies

$$u = \mathcal{F}(u, \lambda) \tag{42}$$

where

$$\mathcal{F}(u, \lambda)(r) = \int_r^R \phi^{-1}\left(\frac{1}{s^{N-1}} \int_0^s \xi^{N-1}(\lambda\psi(u(\xi)) + f(\xi, u(\xi), \lambda)d\xi\right)ds. \tag{43}$$

It is clear that $\mathcal{F} : C[0, R] \times \mathbb{R} \to C[0, R]$ is a completely continuous operator. We have:

Theorem 4.1 *(i) Suppose that $f : [0, R] \times \mathbb{R} \times \mathbb{R}$ is continuous and satisfies $f(r, s, \lambda) = o(|\psi(s)|)$ near $s = 0$ uniformly for r and λ in bounded intervals, and that ϕ and ψ satisfy (25). If $(0, \overline{\lambda})$ is a bifurcation point for (42), then $\overline{\lambda} = \lambda_n(p)$, for some $n \in \mathbb{N}$.*
(ii) Suppose that $f : [0, R] \times \mathbb{R} \times \mathbb{R}$ is continuous and satisfies $f(r, s, \lambda) = o(|\psi(s)|)$ near infinity, uniformly for r and λ in bounded intervals, and that ϕ and ψ satisfy (25). If $(\infty, \overline{\lambda})$ is a bifurcation point for (42), then $\overline{\lambda} = \lambda_n(q)$, for some $n \in \mathbb{N}$.

Proof We prove only *(i)*. The proof of *(ii)* is similar. Suppose that $(0, \overline{\lambda})$ is a bifurcation point for (42), then there exists a sequence $\{(u_n, \lambda_n)\}$ in $C[0, R] \times \mathbb{R}$ with $(u_n, \lambda_n) \to (0, \overline{\lambda})$ and such that (u_n, λ_n) satisfies (42) for each $n \in \mathbb{N}$. Equivalently, (u_n, λ_n) satisfies

$$u_n(r) = \int_r^R \phi^{-1}\left(\frac{1}{s^{N-1}} \int_0^s \xi^{N-1}(\lambda_n\psi(u_n(\xi)) + f(\xi, u_n(\xi), \lambda_n)d\xi\right)ds. \tag{44}$$

Let us set $\varepsilon_n = \|u_n\|$ and $\hat{u}_n(r) = \frac{u_n(r)}{\varepsilon_n}$, then

$$\hat{u}_n(r) = \frac{\int_r^R \phi^{-1}\left(\frac{1}{s^{N-1}} \int_0^s \xi^{N-1}(\lambda_n\psi(u_n(\xi)) + f(\xi, u_n(\xi), \lambda_n)d\xi\right)ds}{\phi^{-1}(\phi(\varepsilon_n)))}, \tag{45}$$

and

$$\hat{u}'_n(r) = \frac{\phi^{-1}\left(\frac{1}{r^{N-1}} \int_0^r \xi^{N-1}\left[\lambda_n\psi(u_n(\xi)) + f(\xi, u_n(\xi), \lambda_n)\right]d\xi\right)}{\phi^{-1}(\phi(\varepsilon_n))}. \tag{46}$$

Using that ϕ is increasing and that $f(r, s, \lambda) = o(|\psi(s)|)$ uniformly for r and λ in compact intervals, we find that the term in the square bracket at the right hand side of (46) is bounded and hence that $\hat{u}_n'$ satisfies

$$||\hat{u}_n'|| \leq \frac{\phi^{-1}(\psi(C\varepsilon_n))}{\phi^{-1}(\psi(\varepsilon_n))} \frac{\phi^{-1}(\psi(\varepsilon_n))}{\phi^{-1}(\phi(\varepsilon_n))}. \tag{47}$$

Since ϕ, ψ satisfy (25) we have that the sequence $\{\hat{u}_n'\}$ is bounded and thus that the sequence $\{\hat{u}_n\}$ is equicontinuous. We thus may assume, without loss in generality that $\hat{u}_n \to \hat{u}$; of course $||\hat{u}|| = 1$.
Writing (45) as

$$\hat{u}_n(r) = \frac{\int_r^R \phi^{-1}(\phi(\varepsilon_n) h_n(s))}{\phi^{-1}(\phi(\varepsilon_n))}, \tag{48}$$

where

$$h_n(s) = \frac{1}{s^{N-1}} \int_0^s \xi^{N-1} \frac{(\lambda_n \psi(u_n(\xi)) + f(\xi, u_n(\xi), \lambda_n))d\xi}{\phi(\varepsilon_n)}, \tag{49}$$

it follows from the Lebesgue dominated convergence theorem applied to (49) that $h_n \to h$, where

$$h(s) = \frac{\overline{\lambda}}{s^{N-1}} \int_0^s \xi^{N-1} \phi_p(\hat{u}(\xi))d\xi. \tag{50}$$

Applying then the Lebesgue dominated convergence theorem to (48), we find that

$$\hat{u}(r) = \int_r^R \phi_p^{-1}(h(s))ds,$$

i.e.

$$\hat{u}(r) = \int_r^R \phi_p^{-1}\Big[\frac{\overline{\lambda}}{s^{N-1}} \int_0^s \xi^{N-1} \phi_p(\hat{u}(\xi))d\xi\Big]ds,$$

which by differentiation implies that $\hat{u}$ is a solution of

$$(r^{N-1}\phi_p(\hat{u}'(r))' + \overline{\lambda} r^{N-1} \phi_p(\hat{u}(r)) = 0$$
$$\hat{u}(R) = 0 = \hat{u}'(0).$$

Thus $\overline{\lambda}$ must be an eigenvalue of the p−Laplacian operator.

4.1   Subcritical problems

We next study the existence of positive solutions to the problem

$$\left[r^{N-1}\phi(u')\right]' + r^{N-1}g(u) = 0, \ r \in (0, R),$$
$$u'(0) = 0, \ \ u(R) = 0, \tag{51}$$

where ϕ is an odd increasing homeomorphism of $\mathbb{R}$ and $g \in C(\mathbb{R})$ is such that $g(x) \geq 0$ for all $x \geq 0$ and $g(0) = 0$. (We ignore here the dependence upon λ since it is of no significance for equations with superlinear but subcritical growth (with respect to ϕ.). Motivated by the case of the p-Laplacian operator and related work we will assume that ϕ satisfies:

$$\limsup_{x \to +\infty} \frac{\phi(\sigma x)}{\phi(x)} < +\infty \tag{52}$$

and

$$\limsup_{x \to 0} \frac{\phi(\sigma x)}{\phi(x)} < +\infty, \tag{53}$$

for all $\sigma > 1$. Also, we will set

$$\Phi(x) = \int_0^x \phi(s)ds, \quad G(x) = \int_0^x g(s)ds, \quad \hat{G}(x) = \int_0^x \hat{g}(s)ds,$$

where $\hat{g}(x) = \sup_{s \in [0,x]} g(s)$. Furthermore we will denote by

$$\Gamma := \limsup_{x \to +\infty} \frac{\Phi(x)}{x\phi(x)},$$

and for $\theta \in (0, 1)$,

$$\delta_\theta := \liminf_{x \to +\infty} \frac{G(\theta x)}{x\hat{g}(x)}.$$

Our main result in this section is the following:

Theorem 4.2 *Suppose that $\Gamma < 1$ and*

$$\lim_{x \to 0} \frac{\phi(x)}{g(x)} = +\infty, \quad \lim_{x \to +\infty} \frac{g(x)}{\phi(x)} = +\infty. \tag{54}$$

Further let there exist $\theta \in (0, 1)$ such that $\delta_\theta > 0$ and

$$N\delta_\theta > N\Gamma - 1. \tag{55}$$

Then problem (51) has at least one positive solution.

Condition (55) has been used to indicate subcritical growth of g respect to ϕ. In the case of powers, i.e., when $\phi(x) = |x|^{p-2}x$ and $g(x) = |x|^{\delta-1}x$, $1 < p < N$ and $\delta > 0$, this condition reads

$$\delta < \frac{N(p-1)+p}{N-p},$$

which is known to be optimal in the sense that there are no positive solutions to problem (51) if

$$\delta \geq \frac{N(p-1)+p}{N-p}. \tag{56}$$

In the case of general ϕ and g, nonexistence of positive solutions to (51) can be proved under the assumption

$$\sup_{x\geq 0} \frac{NG(x)}{xg(x)} \leq \inf_{x\geq 0} \frac{N\Phi(x)}{x\phi(x)} - 1 \tag{57}$$

(see e.g. [30]).

In proving the above theorem, we use techniques similar to those used in the previous section. We note that u be a solution of (51) if and only if $u(r)$ satisfies the integral equation

$$u(r) = \int_r^R \phi^{-1}\left[\frac{1}{s^{N-1}} \int_0^s \xi^{N-1}g(|u(\xi)|)d\xi\right]ds. \tag{58}$$

Let us now define $T_0 : C_\# \to C_\#$ by

$$T_0(u)(r) = \int_r^R \phi^{-1}\left[\frac{1}{s^{N-1}} \int_0^s \xi^{N-1}g(|u(\xi)|)d\xi\right]ds. \tag{59}$$

Clearly T_0 is well defined and fixed points of T_0 will provide solutions of (51). Define now the operator $T : C_\# \times [0,1] \to C_\#$, by

$$T(u,\lambda)(r) = \int_r^R \phi^{-1}\left[\frac{1}{s^{N-1}} \int_0^s \xi^{N-1}(g(|u(\xi)|) + \lambda h)d\xi\right]ds, \tag{60}$$

where $h > 0$ is a constant to be fixed later. We find that T sends bounded sets of $C_\# \times [0,1]$ into bounded sets of $C_\#$ and $T(u,0) = T_0(u)$. Moreover, T is a completely continuous operator.

Also, let us define the operator $S : C_\# \times [0,1] \to C_\#$,

$$S(u,\lambda) = \int_r^R \phi^{-1}\left[\frac{\lambda}{s^{N-1}} \int_0^s \xi^{N-1}g(|u(\xi)|)d\xi\right]ds. \tag{61}$$

Again, we see that S is completely continuous and note that $S(\cdot,1) = T_0$.

We will prove the existence of a fixed point of T_0, and hence of a positive solution to (51), by using suitable a-priori estimates and degree theory. Indeed, we will show that there exist $R_1 > 0$ and $\varepsilon_0 > 0$ such that

$$\deg_{LS}(I - T_0, B(0,R_1), 0) = 0 \tag{62}$$

and

$$\deg_{LS}(I - T_0, B(0, \varepsilon_0), 0) = 1 \qquad (63)$$

and thus, using the excision property of the Leray-Schauder degree it will follow that there must exist a solution of the equation

$$u = T_0(u)$$

in $B(0, R_1) \setminus \overline{B(0, \varepsilon_0)}$.

To compute the degree in (62) we shall use the operator T defined by (60) and note that fixed points of T are solutions of

$$\left[r^{N-1}\phi(u')\right]' + r^{N-1}(g(|u|) + h) = 0, \ r \in (0, R),$$
$$u'(0) = 0, \quad u(R) = 0. \qquad (64)$$

We have the following lemma:

Lemma 4.1 *There exists $h_0 > 0$ such that problem (64) has no solutions for $h \geq h_0$.*

Proof We argue by contradiction and thus assume that there exists a sequence $\{h_n\}_{n \in \mathbb{N}}$, with $h_n \to +\infty$ as $n \to \infty$ such that problem (64) has a solution u_n for each $n \in \mathbb{N}$. Then, u_n satisfies

$$-r^{N-1}\phi(u_n'(r)) = \int_0^r \xi^{N-1}(g(|u_n(\xi)|) + h_n)d\xi \geq 0$$

and

$$u_n(r) = \int_r^R \phi^{-1}\left[\frac{1}{s^{N-1}} \int_0^s \xi^{N-1}(g(|u_n(\xi)|) + h_n)d\xi\right] ds \geq 0.$$

Hence, $u_n'(r) \leq 0$ and $u_n(r) \geq 0$ for all $r \in [0, R]$. Also,

$$u_n(r) \geq (R - r)\phi^{-1}\left(\frac{rh_n}{N}\right), \quad \text{for all} \quad r \in [0, R]$$

and thus, if $r \in \left[\frac{R}{4}, \frac{3R}{4}\right]$,

$$u_n(r) \geq \frac{R}{4}\phi^{-1}\left(\frac{Rh_n}{4N}\right). \qquad (65)$$

Since we also have that for $r \in \left[\frac{R}{4}, \frac{3R}{4}\right]$

$$u_n(r) \geq \int_r^{\frac{3R}{4}} \phi^{-1}\left[\frac{1}{s^{N-1}} \int_0^s \xi^{N-1}g(|u_n(\xi)|)d\xi\right] ds,$$

by (65), the fact that u_n is decreasing, and the second assumption in (53), we have, that given $A > 0$, there exists $n_1 \in \mathbb{N}$ such that for $n \geq n_1$

$$u_n(r) \geq \int_r^{\frac{3R}{4}} \phi^{-1}\left[\frac{A}{s^{N-1}} \int_0^s \xi^{N-1} \phi(u_n(\xi)) d\xi\right] ds$$

$$\geq \int_r^{\frac{3R}{4}} \phi^{-1}\left[\frac{A}{s^{N-1}} \int_0^r \xi^{N-1} \phi(u_n(\xi)) d\xi\right] ds,$$

and since $\phi(u_n(\cdot))$ is decreasing,

$$u_n(r) \geq \int_r^{\frac{3R}{4}} \phi^{-1}\left[\frac{Ar^N}{Ns^{N-1}} \phi(u_n(r))\right] ds.$$

Using now that $\frac{R}{4} \leq r \leq s \leq \frac{3R}{4}$, we obtain

$$\frac{Ar^N}{Ns^{N-1}} \geq dA,$$

where d is a constant. Thus, by the monotonicity of ϕ^{-1}, for all $r \in \left[\frac{R}{4}, \frac{R}{2}\right]$, we have that

$$u_n(r) \geq \frac{R}{4} \phi^{-1}(dA\phi(u_n(r))), \quad \text{for all} \quad n \geq n_1,$$

or equivalently,

$$\frac{\phi\left(\frac{4}{R} u_n(r)\right)}{\phi(u_n(r))} \geq dA, \tag{66}$$

for all $n \geq n_1$ and for all $r \in \left[\frac{R}{4}, \frac{R}{2}\right]$.

It is clear that (66) cannot hold for A large if $R \geq 4$. If $R < 4$, by (65), condition (52), and the fact that A is arbitrary, we obtain a contradiction for $n \geq n_0$ for some $n_0 \geq n_1$.

Let us fix now $h \geq h_0$ for h_0 given in Lemma 4.1, and consider the family of problems

$$\begin{align} &\left[r^{N-1}\phi(u')\right]' + r^{N-1}(g(|u|) + \lambda h) = 0, \ r \in (0, R), \\ &u'(0) = 0, \quad u(R) = 0, \quad \lambda \in [0, 1]. \end{align} \tag{67}$$

Lemma 4.2 *Solutions to (67) are a-priori bounded.*

The proof of Lemma 4.2 is lengthy and depends on certain energy estimates. We shall not give the proof here but refer to [30].

We need one further lemma to do the necessary degree computations. Its proof follows by arguments similar to those used above.

Lemma 4.3 *There exists an $\varepsilon_0 > 0$ such that the equation*

$$u = S(u, \lambda) \tag{68}$$

has no solution (u, λ) with $u \in \partial B(0, \varepsilon_0)$ and $\lambda \in [0, 1]$.

It follows from Lemmas 4.1 and 4.2, that if u is a solution to the equation

$$u = T(u, \lambda), \quad \lambda \in [0, 1],$$

then $\|u\|_\infty \leq C$, where C is a positive constant. Thus if $B(0, R_1)$ denotes the ball centered at 0 in $C_\#$ with radius $R_1 > C$, we have that the Leray-Schauder degree of the operator

$$I - T(\cdot, \lambda) : \overline{B(0, R_1)} \rightarrow C_\#$$

is well defined for every $\lambda \in [0, 1]$. Then, by the properties of the Leray-Schauder degree, we have that

$$\deg_{LS}(I - T_0, B(0, R_1), 0) = \deg_{LS}(I - T(\cdot, 1), B(0, R_1), 0) = 0, \tag{69}$$

viz. Lemma 4.1. Also, by Lemma 4.2 and the properties of the Leray Schauder degree, it follows that for $\varepsilon_0 > 0$ small enough,

$$\deg_{LS}(I - S(\cdot, \lambda), B(0, \varepsilon_0), 0) = \text{constant}, \quad \text{for all} \quad \lambda \in [0, 1].$$

Hence

$$\deg_{LS}(I - T_0, B(0, \varepsilon_0), 0) = \deg_{LS}(I, B(0, \varepsilon_0), 0) = 1.$$

Thus, using the excision property of the Leray-Schauder degree we conclude that there must be a solution of the equation

$$u = T_0(u)$$

with $u \in B(0, R_1) \setminus \overline{B(0, \varepsilon_0)}$, completing the proof of Theorem 4.2.

5 Problems on annular domains

In this section we consider nonlinear ordinary differential equations of the form

$$\begin{cases} (\phi(u'))' + \frac{N-1}{r}\phi(u') + f(u) = 0, \ 0 < r_1 < r < r_2 \\ u = 0, \ r \in \{r_1, r_2\}, \end{cases} \tag{70}$$

where $\phi : \mathbb{R} \rightarrow \mathbb{R}$ is an odd increasing homeomorphism and which satisfies:

$$\forall c > 0, \ \exists A_c > 0,$$

such that

$$A_c \phi(u) \leq \phi(cu), \ u \in \mathbb{R}^+, \tag{71}$$

where

$$\lim_{c \to \infty} A_c = \infty \tag{72}$$

Note that the above assumption immediately implies that $\exists B_c > 0$ such that

$$\phi(cu) \leq B_c\phi(u), \ u \in \mathbb{R}^+,$$

with

$$\lim_{c \to 0} B_c = 0.$$

Since the only property of the function $\frac{N-1}{r}$ we shall use here is its continuity, we consider the more general problem

$$\begin{cases} (\phi(u'))' + b(r)\phi(u') + f(u) = 0, \ 0 < r_1 < r < r_2 \\ u = 0, \ r \in \{r_1, r_2\}, \end{cases} \tag{73}$$

where $b : [r_1, r_2] \to \mathbb{R}$ is a continuous function. The nonlinear term f will be assumed to be continuous and satisfy

$$\lim_{u \to 0} \frac{f(u)}{\phi(u)} \leq 0 \tag{74}$$

and to grow superlinearly (with respect to ϕ) near infinity, i.e.

$$\lim_{u \to \infty} \frac{f(u)}{\phi(u)} = \infty. \tag{75}$$

For such problems we shall establish that (73) always has a positive solution defined for any interval $[r_1, r_2]$. The results obtained may be viewed as extensions of results in [16], [38] to p-Laplacian like equations, and of results in [34], [40], [51], and [53]. The results obtained will be equally valid in case f depends upon r also, provided the assumptions made are assumed uniform with respect to r.

5.1 Fixed point formulation

Letting

$$p(r) = e^{\int_{r_1}^{r} b},$$

we may rewrite problem (73) equivalently as

$$\begin{cases} (p\phi(u'))' + pf(u) = 0, \ 0 < r_1 < r < r_2 \\ u = 0, \ r \in \{r_1, r_2\}. \end{cases} \tag{76}$$

We shall establish the existence of solutions of (76) (hence (73) by proving the existence of fixed points of a completely continuous operator F,

$$F : C[r_1, r_2] := E \to E,$$

where the norm of E is given by $\|u\| = \max_{r \in [r_1, r_2]} |u(r)|$. The operator F is defined by the following lemma.

Lemma 5.1 *Let ϕ be as above and c a nonnegative constant. Then for each $v \in E$ the problem*

$$\begin{cases} (p\phi(u'))' - cp\phi(u) = pv, \ 0 < r_1 < r < r_2 \\ u = 0, \ r \in \{r_1, r_2\} \end{cases} \tag{77}$$

has a unique solution

$$u = T(v),$$

and the operator $T : E \to E$ is completely continuous.

Proof For each $w \in E$ let $u = B(w)$ be the unique solution of

$$\begin{cases} (p\phi(u'))' - cp\phi(w) = pv, \ 0 < r_1 < r < r_2 \\ u = 0, \ r \in \{r_1, r_2\}, \end{cases}$$

i.e. u is given by

$$u(r) = \int_{r_1}^{r} \phi^{-1}\left(\frac{1}{p}\{q - \int_{r_1}^{s}(p(c\phi(w) + v)\}\right) ds,$$

where q is the unique number such that $u(r_2) = 0$.

From this also follows that B is a completely continuous mapping.

We shall next employ the continuation theorem of Leray-Schauder to establish that the operator B has a fixed point u, i.e. that

$$\begin{cases} (p\phi(u'))' - cp\phi(u) = pv, \ 0 < r_1 < r < r_2 \\ u = 0, \ r \in \{r_1, r_2\}, \end{cases} \tag{78}$$

has a solution. We then define the operator T by

$$T(v) = u,$$

where u is the solution of (77). To accomplish what has been said, let $u \in E$ and $\lambda \in (0,1)$ be such that

$$u = \lambda B(u).$$

Then

$$\begin{cases} \left(p\phi\left(\frac{u'}{\lambda}\right)\right)' - cp\phi(u) = pv, \ 0 < r_1 < r < r_2 \\ u = 0, \ r \in \{r_1, r_2\}. \end{cases}$$

Multiplying the above by u and integrating we obtain

$$\int_{r_1}^{r_2} p\phi\left(\frac{u'}{\lambda}\right) u' + \int_{r_1}^{r_2} cp\phi(u)u = -\int_{r_1}^{r_2} pvu.$$

On the other hand, since ϕ is an increasing homeomorphism, for each $\epsilon > 0$, there exists a constant c_ϵ such that

$$|x| \le \epsilon\phi(x)x + c_\epsilon, \ x \in \mathbb{R}.$$

Thus, we obtain

$$\left| \int_{r_1}^{r_2} pvu \right| \leq \|v\| \int_{r_1}^{r_2} p|u| \leq \|v\| \left\{ \epsilon \int_{r_1}^{r_2} p\phi(u)u + c_\epsilon(r_2 - r_1) \right\},$$

and choosing ϵ appropriately

$$\left| \int_{r_1}^{r_2} pvu \right| \leq \frac{1}{2}c \int_{r_1}^{r_2} p\phi(u)u + c_1,$$

where c_1 is a constant. Hence we obtain

$$\int_{r_1}^{r_2} p\phi\left(\frac{u'}{\lambda}\right) u' \leq c_2,$$

for a constant c_2. Therefore

$$\int_{r_1}^{r_2} |u'| \leq c_3,$$

and hence

$$\|u\| \leq c_4.$$

Further

$$\left\| \left(p\phi\left(\frac{u'}{\lambda}\right) \right)' \right\|_{L^1} \leq c_5.$$

Since there exists r_0 such that $u'(r_0) = 0$, we obtain from the latter inequality that

$$\left| p\phi\left(\frac{u'}{\lambda}\right) \right| \leq c_6$$

and hence

$$\|u'\| \leq c_7,$$

where $c_1, \cdots, c_7$ are constants independent of λ. The complete continuity of B is easily established and we conclude that B has a fixed point. If u_1 and u_2 are fixed points, one immediately obtains that

$$\int_{r_1}^{r_2} p(\phi(u_1') - \phi(u_2'))(u_1' - u_2') + \int_{r_1}^{r_2} cp(\phi(u_1) - \phi(u_2))(u_1 - u_2) = 0,$$

and hence, since ϕ is increasing, that $u_1 = u_2$. Thus the operator T given in the statement of the lemma is well defined. That it is completely continuous, again may easily be established.

Using the Lemma 5.1 we may obtain a fixed point formulation of problem (76) as

$$u = F(u) = T(-c\phi(u) - f(u)). \qquad (79)$$

The next lemma is crucial for establishing the existence of nonzero fixed points of (79).

Lemma 5.2 *If u is the solution of (78) with $v \leq 0$, then*

$$u(r) \geq c_0 \|u\| k(r), \ r_1 \leq r \leq r_2,$$

where

$$k(r) = \frac{1}{r_2 - r_1} \min\{r - r_1, r_2 - r\},$$

and c_0 is a positive constant.

Proof Let r_0 be such that $\|u\| = |u(r_0)|$. Let w be the solution of

$$\begin{cases} (p\phi(w'))' - cp\phi(w) = 0, \ 0 < r_1 < r < r_2 \\ w(r_1) = 0, \ w(r_0) = \|u\|, \end{cases}$$

An argument, similar to the uniqueness argument used above, shows that $u \geq 0, r_1 \leq r \leq r_2$, $w \geq 0, r_1 \leq r \leq r_0$, and $u - w \geq 0, r_1 \leq r \leq r_0$. Further $w' \geq 0$, $r_1 \leq r \leq r_0$. Next, observe

$$p(r)\phi(w'(r)) = \phi(w'(r_1)) + c \int_{r_1}^{r} p\phi(w),$$

which implies

$$\phi(w'(r)) \leq c_1 \phi(w'(r_1)) + c_2 \int_{r_1}^{r} \phi \left(\int_{r_1}^{s} w' \right).$$

Using (72), we get

$$\int_{r_1}^{s} w' = \int_{r_1}^{s} \phi^{-1} \left\{ \frac{p\phi(w')}{p} \right\} \leq \int_{r_1}^{s} \phi^{-1}(c_3 p\phi(w'))$$
$$\leq c_4 \int_{r_1}^{s} \phi^{-1}(p\phi(w')) \leq c_5 \phi^{-1}(p\phi(w'(s))),$$

and therefore

$$\phi \left(\int_{r_1}^{s} w' \right) \leq c_6 p\phi(w'(s)) \leq c_7 \phi(w'(s)).$$

We therefore obtain

$$\phi(w'(r)) \leq c_1 \phi(w'(r_1)) + c_2 c_7 \int_{r_1}^{r} \phi(w'),$$

and hence by Gronwall's inequality

$$\phi(w'(r)) \leq c_1 \phi(w'(r_1)) e^{c_2 c_7 (r - r_1)} \leq c_8 \phi(w'(r_1)),$$

or

$$w'(r) \leq c_9 w'(r_1).$$

Integrating, we obtain

$$\|u\| \leq c_9 (r_2 - r_1) w'(r_1).$$

We hence find

$$u(r) \geq w(r) = \int_{r_1}^r w' = \int_{r_1}^r \phi^{-1}\left\{\frac{p\phi(w')}{p}\right\}$$
$$\geq c_{10} \int_{r_1}^r \phi^{-1}(p\phi(w')) \geq c_{10}(r - r_1)\phi^{-1}(p(r_1)\phi(w'(r_1)))$$
$$\geq c_{11}(r - r_1)\|u\|, \ r_1 \leq r \leq r_0.$$

Using a similar argument, we see that

$$u(r) \geq c_{12}(r_2 - r)\|u\|, \ r_0 \leq r \leq r_2,$$

where in the above calculations $c_1, \cdots, c_{12}$ are constants that only depend upon p, ϕ and the length of the interval.

5.2 Existence results

We next establish the existence of a nontrivial solution of (76) by means of a fixed point argument (a cone expansion theorem analogous to results in [43] and [38]) and subsequently impose conditions upon f which will allow us to use this result.

We first look at some auxiliary results.

Proposition 5.1 *Assume that* $f : \mathbb{R} \to \mathbb{R}$ *is continuous and there exists a constant* $c > 0$ *such that*

$$f(u) + c\phi(u) \geq 0, \ u \geq 0.$$

Further assume there exists a constant $m > 0$ *such that*

$$u = T(\lambda(-c\phi(|u|) - f(|u|)), \ 0 \leq \lambda \leq 1, \ \Rightarrow \|u\| \neq m,$$

and there exists a constant $M > m$ *and an element* $h \in E$, $h \leq 0$ *such that*

$$u = T(-c\phi(u) - f(|u|) + \lambda h), \ 0 \leq \lambda \leq 1, \ \Rightarrow \|u\| \neq M,$$

further any solution u *of*

$$u = T(-c\phi(u) - f(|u|) + h),$$

satisfies $\|u\| > M$. *Then there exists a fixed point* $u \in E$ *of the operator* F, $F(u) = T(-c\phi(|u|) - f(|u|))$ *such that*
$$m < \|u\| < M.$$

Rephrasing Proposition 5.1 for the boundary value problem (76), we have the following corollary.

Corollary 5.1 *Assume that* $f : \mathbb{R} \to \mathbb{R}$ *is continuous and there exists a constant* $c > 0$ *such that*

$$f(u) + c\phi(u) \geq 0, \ u \geq 0.$$

Further assume there exists a constant $m > 0$ *such that for* $0 \leq \lambda \leq 1$ *and any solution* u *of*

$$\begin{cases} (p\phi(u'))' - cp\phi(u) + \lambda(cp\phi(|u|) + pf(|u|)) = 0, \ 0 < r_1 < r < r_2, \\ u = 0, \ r \in \{r_1, r_2\}. \end{cases} \tag{80}$$

satisfies

$$\|u\| \neq m,$$

and there exists a constant $M > m$ *and an element* $h \in E$, $h \geq 0$ *such that any solution* u *of*

$$\begin{cases} (p\phi(u'))' - p\phi(u) + p\phi(|u|) + pf(|u|) \leq 0, \ 0 < r_1 < r < r_2, \\ u = 0, \ r \in \{r_1, r_2\}. \end{cases} \tag{81}$$

satisfies $\|u\| \neq M$, *and solutions* u *of*

$$\begin{cases} (p\phi(u'))' - p\phi(u) + p\phi(|u|) + pf(|u|) + ph = 0, \ 0 < r_1 < r < r_2, \\ u = 0, \ r \in \{r_1, r_2\}, \end{cases} \tag{82}$$

satisfy $\|u\| > M$. *Then there exists a solution* u, $u \geq 0$ *of (43) such that*

$$m < \|u\| < M.$$

We next provide conditions on f and further conditions on ϕ in order that the above corollary may be applied.

To obtain our main result we provide conditions about the behavior of f near zero and near infinity. These conditions in turn will provide the validity of Corollary 5.1 and hence yield the existence of nontrivial solutions.

Proposition 5.2 *Assume that* $f : \mathbb{R} \to \mathbb{R}$ *is continuous and let there exist a constant* $c \geq 0$ *such that*

$$f(u) + c\phi(u) \geq 0, \ u \geq 0.$$

Further, let ϕ *be an odd increasing homeomorphism of* $\mathbb{R}$ *which satisfies (71) and assume (74) holds. Then there exists a positive number* m *such that for* $0 \leq \lambda \leq 1$ *and any solution* u *of*

$$\begin{cases} (p\phi(u'))' - cp\phi(u) + \lambda(pf(|u|) + cp\phi(|u|)) = 0, \ 0 < r_1 < r < r_2, \\ u = 0, \ r \in \{r_1, r_2\} \end{cases}$$

satisfies $\|u\| \neq m$.

Proof We observe that earlier considerations imply that solutions are nonnegative. To prove the result we argue indirectly and conclude the existence of sequences $\{\lambda_n\} \subset [0, 1]$, $\{\epsilon_n\}$, $\epsilon_n \searrow 0$, $\{u_n\}$, $\|u_n\| = \epsilon_n$, such that the triple $(\epsilon_n, \lambda_n, u_n)$ is a solution. Since each $u_n \neq 0$ has a maximum point, say $u_n(r_n^*) = \|u_n\|$, we obtain by integration

$$u_n'(r) \leq \phi^{-1}\left(\frac{\lambda_n}{p(r)} \int_r^{r_n^*} p(s)f(u_n(s))ds\right);$$

and hence, for given $\epsilon > 0$ we have for large n

$$u_n'(r) \leq \phi^{-1}\left(\frac{\lambda_n}{p(r)} \int_r^{r_n^*} \epsilon p(s)\phi(u_n(s))ds\right), \quad r \leq r_n^*;$$

integrating once more and using the fact that the integrand is positive and that ϕ^{-1} is monotone, we obtain

$$\|u_n\| \leq \int_{r_1}^{r_2} \left(\phi^{-1}\left(\frac{1}{p(r)} \int_{r_1}^{r_2} \epsilon p(s)\phi(u_n(s))ds\right)\right)dt;$$

hence for all large n

$$\|u_n\| \leq (r_2 - r_1)\phi^{-1}(c_1\epsilon\phi(\epsilon_n)),$$

or

$$\phi\left(\frac{\epsilon_n}{r_2 - r_1}\right) \leq c_1\epsilon\phi(\epsilon_n)),$$

hence if $r_2 - r_1 \leq 1$ we immediately obtain a contradiction for ϵ small, on the other hand if if $r_2 - r_1 > 1$, we obtain a contradiction using (72).

We next consider conditions which guarantee the validity of the second part of Corollary 5.1.

Proposition 5.3 *Assume that $f : \mathbb{R} \to \mathbb{R}$ is continuous and let there exist a constant $c \geq 0$ such that*

$$f(u) + c\phi(u) \geq 0, \ u \geq 0.$$

Further, let ϕ be an odd increasing homeomorphism of $\mathbb{R}$ which satisfies (71); also assume (75) holds. Then there exists a positive number M such that any solution u of

$$\begin{cases} (p\phi(u'))' - cp\phi(u) + cp\phi(|u|) + pf(|u|) \leq 0, \ 0 < r_1 < r < r_2, \\ u = 0, \ r \in \{r_1, r_2\} \end{cases}$$

satisfies $\|u\| \neq M$.

Proof Let u be a solution with $|u(r_0)| = \|u\|$. It follows from earlier considerations that u must be nonnegative. Hence, integrating, we obtain

$$u'(r) \geq \phi^{-1}\left(\frac{1}{p(r)} \int_{r_0}^r p(s)f(u(s))ds\right), \quad r \leq r_0.$$

By a new integration, we conclude

$$u(\tau) \geq \int_{r_1}^{\tau} \left(\phi^{-1} \left(\frac{1}{p(r)} \int_r^{r_0} p(s) f(u(s)) ds \right) \right) dr, \ \tau \leq r_0.$$

In the above inequality we now add and subtract the term

$$\frac{1}{p(r)} \int_r^{r_0} cp(s) \phi(u(s)) ds$$

under the integral.

If then $r_0 \geq \frac{r_1+r_2}{2}$, the above inequality yields

$$\|u\| \geq \left(\frac{r_2 - r_1}{3} \right) \phi^{-1} \left(c_1 \int_{\frac{2r_1+r_2}{3}}^{\frac{r_2+r_1}{2}} (f(u) + c\phi(u)) - c_2 \int_{r_1}^{r_2} \phi(u) \right),$$

or

$$\phi \left(\frac{3}{r_2 - r_1} \|u\| \right) \geq c_1 \phi(\delta \|u\|) \int_{\frac{2r_1+r_2}{3}}^{\frac{r_2+r_1}{2}} \frac{f(u) + c\phi(u)}{\phi(u)} - c_2 (r_2 - r_1) \phi(\|u\|),$$

where $\delta = c_0 \min\{k(r) : \frac{2r_1+r_2}{3} \leq r \leq \frac{2r_2+r_1}{3}\}$, and c_1, c_2 are positive constants. If, on the other hand, $r_0 \leq \frac{r_1+r_2}{2}$, we obtain, via similar considerations

$$\phi \left(\frac{3}{r_2 - r_1} \|u\| \right) \geq c_1 \phi(\delta \|u\|) \int_{\frac{r_1+r_2}{2}}^{\frac{2r_2+r_1}{3}} \frac{f(u) + c\phi(u)}{\phi(u)} - c_2 (r_2 - r_1) \phi(\|u\|),$$

with δ, c_1, c_2 as above. Combining the two cases we obtain

$$\phi \left(\frac{3}{r_2 - r_1} \|u\| \right) \geq (c_3 I(u) - c_2 (r_2 - r_1)) \phi(\|u\|),$$

where

$$I(u) = \min \left\{ \int_{\frac{r_1+r_2}{2}}^{\frac{2r_2+r_1}{3}} \frac{f(u) + c\phi(u)}{\phi(u)}, \int_{\frac{2r_1+r_2}{3}}^{\frac{r_2+r_1}{2}} \frac{f(u) + c\phi(u)}{\phi(u)} \right\}.$$

On the other hand

$$\lim_{\|u\| \to \infty} I(u) = \infty.$$

The conclusion therefore follows from the properties of ϕ.

We need one further result to be able to apply Corollary 5.1.

Proposition 5.4 *Assume the hypotheses of Proposition 5.3 and let M be a constant whose existence is guaranteed there. Then there exists $h \in E, h \geq 0$ such that $\|u\| > M$ for any solution of*

$$\begin{cases} (p\phi(u'))' - cp\phi(u) + cp\phi(|u|) + pf(|u|) + ph = 0, \ 0 < r_1 < r < r_2, \\ u = 0, \ r \in \{r_1, r_2\}. \end{cases}$$

Proof If u is a solution with $|u(r_0)| = \|u\|$, we may proceed as in the proof of the previous proposition and conclude that, if $r_0 \geq \frac{r_1 + r_2}{2}$, then

$$\|u\| \geq \left(\frac{r_2 - r_1}{3}\right) \phi^{-1} \left(c_1 \int_{\frac{2r_1 + r_2}{3}}^{\frac{r_2 + r_1}{2}} h - c_2 \int_{r_1}^{r_2} \phi(u)\right).$$

If, on the other hand, $r_0 \leq \frac{r_1 + r_2}{2}$, we obtain, via similar considerations

$$\phi\left(\frac{3}{r_2 - r_1}\|u\|\right) \geq c_1 \int_{\frac{r_1 + r_2}{2}}^{\frac{2r_2 + r_1}{3}} h - c_2(r_2 - r_1)\phi(\|u\|).$$

Combining the two cases we obtain

$$\phi\left(\frac{3}{r_2 - r_1}\|u\|\right) \geq c_3 I - c_2(r_2 - r_1))\phi(\|u\|),$$

where

$$I = \min\left\{\int_{\frac{r_1 + r_2}{2}}^{\frac{2r_2 + r_1}{3}} h, \int_{\frac{2r_1 + r_2}{3}}^{\frac{r_2 + r_1}{2}} h\right\}.$$

Thus, letting $h = constant \gg 1$, we obtain the result.

Combining the above propositions and using Corollary 5.1 we have proved our main result.

Theorem 5.1 *Assume that $f : \mathbb{R} \to \mathbb{R}$ is continuous. Further, let ϕ be an odd increasing homeomorphism of $\mathbb{R}$ which satisfies (71), and let the following conditions hold:*

$$-\infty < \liminf_{u \to 0} \frac{f(u)}{\phi(u)} \leq \limsup_{u \to 0} \frac{f(u)}{\phi(u)} \leq 0,$$

$$\lim_{u \to \infty} \frac{f(u)}{\phi(u)} = \infty.$$

Then the boundary value problem (73) has a positive solution.

We remark here that the condition that there exist a constant $c \geq 0$ such that

$$f(u) + c\phi(u) \geq 0, \ u \geq 0.$$

follows from the assumptions of the theorem.

The next theorem considers the case that $f : \mathbb{R} \to \mathbb{R}$ is nonnegative. The behavior of f near the origin is then described in terms of the functions

$$G(u) = \int_0^u f, \ \Phi(u) = \int_0^u \phi.$$

We have the following result:

Theorem 5.2 *Assume that $f : \mathbb{R} \to \mathbb{R}$ is continuous and nonnegative. Further, let ϕ be an odd increasing homeomorphism of $\mathbb{R}$ such that $\frac{\phi(u)}{u}$ is nondecreasing on $\mathbb{R}^+$ and satisfies (71); also let the following conditions hold:*

$$\liminf_{u \to 0} \frac{G(u)}{\Phi(u)} = 0,$$

$$\lim_{u \to \infty} \frac{f(u)}{\phi(u)} = \infty.$$

Then the boundary value problem (73) has a positive solution.

We note that in the case $\phi(s) = |s|^{p-2}s$ the conditions above hold if and only if $p \geq 2$.

6 Positone problems

In this section we are interested in the existence and multiplicities of positive solutions of the boundary value problem.

$$(\phi(u'))' + \lambda f(t, u) = 0, \ a < t < b$$
$$u(a) = 0 = u(b) \tag{83}$$

with f continuous (but not necessarily locally Lipschitz continuous). We make the following assumptions:

$$\phi \text{ is an odd increasing homeomorphism on } \mathbb{R}$$
$$\limsup_{x \to \infty} \frac{\phi(\sigma x)}{\phi(x)} < \infty, \forall \sigma > 0. \tag{84}$$

$$f : [a, b] \times [0, \infty) \to (0, \infty) \text{ is continuous}$$
$$\exists [c, d] \subset (a, b), \ c < d \ \text{such that} \tag{85}$$
$$\lim_{u \to \infty} \frac{f(t,u)}{\phi(u)} = \infty, \quad \text{uniformly for } t \in [c, d].$$

The main result in this section is:

Theorem 6.1 *Let (84) and (85) hold. Then there exists a positive number λ^* such that the problem (83) has at least two positive solutions for $0 < \lambda < \lambda^*$, at least one for $\lambda = \lambda^*$ and none for $\lambda > \lambda^*$*

Note that in the special case where $\phi(u') = u'$, Theorem 6.1 is a classical (see e.g. [16], [61].) Related results for the case $\phi(u') = |u'|^{p-2}u'$ can be found in [20], [40] and the references in these papers.

It is an easy exercise to see that the type of boundary value studied in the previous section may in fact, via a change of variables, be transformed to a problem of type (83).

I proving Theorem 6.1, we shall, in addition to continuation methods also employ upper and lower solution methods. These methods are, of course, standard for semilinear equations (see [60], [27]) and we refer to [6], [11],[18], [13], [14], [55], where the types of theorems for the nonlinear case are presented.

We now prove Theorem 6.1. Since we are interested in nonnegative solutions we shall make the convention that $f(t, u) = f(t, 0)$ if $u < 0$.

We first need a lemma which is a special case of Lemma 5.2.

Lemma 6.1 *Let $v \in C^0[a, b]$ with $v \leq 0$ and let u satisfy*

$$(\phi(u'))' = v$$

$$u(a) = 0 = u(b).$$

Then

$$u(t) \geq \|u\| p(t), \quad t \in [a, b]$$

where

$$p(t) = \frac{\min(t - a, b - t)}{b - a}$$

The next sequence of lemmas will allow us to employ continuation methods by establishing necessary a priori bounds on solutions.

Lemma 6.2 *Suppose that $g : [a, b] \times \mathbb{R}^+ \to \mathbb{R}^+$ is continuous and there exists a positive number M and an interval $[a_1, b_1] \subset (a, b)$ such that*

$$g(t, u) \geq M(\phi(u) + 1), \quad t \in [a_1, a_2], \ u \geq 0.$$

There exists a positive number $M_0 = M_0(\phi, a_1, b_1)$ such that the problem

$$(\phi(u'))' = -g(t, u)$$
$$u(a) = 0 = u(b)$$

has no solution whenever $M \geq M_0$.

Proof Let u be a solution. Then

$$u(t) = \int_a^t \phi^{-1}[c - \int_a^s g(\tau, u)d\tau]ds,$$

where $c = \phi(u'(a))$. Let $\|u\| = u(t_0)$, $t_0 \in [a, b]$. Then $u'(t_0) = 0$ and hence

$$u(t) = \int_a^t \phi^{-1}[\int_s^{t_0} g(\tau, u)d\tau]ds$$

If $t_0 \geq \frac{a_1 + b_1}{2}$, then

$$\|u\| \geq u(a_1) > \int_a^{a_1} \phi^{-1}[M \int_{a_1}^{\frac{a_1 + b_1}{2}} (\phi(u) + 1)]$$
$$> (a_1 - a)\phi^{-1}[M \frac{(b_1 - a_1)}{2}[\phi(\|u\|\delta) + 1]]$$

where
$$\delta = \min_{a_1 \leq t \leq b_1} p(t).$$

This implies
$$\phi\left(\frac{\|u\|}{a_1 - a}\right) > M\frac{(b_1 - a_1)}{2}[\phi(\|u\|\delta) + 1]$$

If $t_0 \leq \frac{a_1 + b_1}{2}$, then since
$$u(t) = \int_t^b \phi^{-1}[\int_{t_0}^s g(\tau, u)d\tau]ds$$

we deduce
$$\phi\left(\frac{\|u\|}{b - b_1}\right) > \frac{M(b_1 - a_1)}{2}[\phi(\|u\|\delta) + 1]$$

Combining the above, we obtain
$$\phi(\gamma\|u\|) > \frac{M(b_1 - a_1)}{2}[\phi(\|u\|\delta) + 1]$$

where $\gamma = \max\left(\frac{1}{b-b_1}, \frac{1}{a_1-a}\right)$.
Consequently,
$$\|u\| > \frac{1}{\gamma}\phi^{-1}\left[\frac{M(b_1 - a_1)}{2}\right]$$

and
$$\frac{\phi(\gamma\|u\|)}{\phi(\delta\|u\|)} > \frac{M}{2}(b_1 - a_1)$$

a contradiction, if M is sufficiently large.

Remark 6.1 *It follows from the proof, that the problem in Lemma 6.2 has no solution u satisfying*
$$g(t, u(t)) \geq M(\phi(u(t)) + 1), \quad t \in [a_1, a_2],$$
if $M \geq M_0$.

These considerations further imply the following result:

Theorem 6.2 *There exists a positive number $\bar{\lambda}$ such that problem (83) has no solution for $\lambda > \bar{\lambda}$.*

It follows immediately from (85) that there exists a constant $\mu > 0$ such that
$$f(t, u) \geq \mu(\phi(u) + 1), \quad u \in \mathbb{R}^+, \quad c \leq t \leq d.$$

Hence the result follows from the previous lemma. We shall also need the the following lemma, whose proof we omit.

Lemma 6.3 *For each $\mu > 0$, there exists a positive constant C_μ such that the problem*

$$(\phi(u'))' = -\lambda\theta f(t, u) - (1 - \theta)M_0(|\phi(u)| + 1)$$
$$u(a) = 0 = u(b)$$

with $\lambda \geq \mu$, $\theta \in [0, 1]$ and M_0 given by Remark 6.1, has no solution satisfying $\|u\| > C_\mu$.

Now, let Λ be the set of all $\lambda > 0$ such that (83) has a solution and let $\lambda^* = \sup\Lambda$.

Lemma 6.4 $0 < \lambda^* < \infty$ *and* $\lambda^* \in \Lambda$.

Proof $u \in C^0[a, b]$ is a solution if and only if $u = F(\lambda, u)$, where

$$F : [0, \infty) \times C^0[a, b] \to C^0[a, b]$$

is the completely continuous mapping given by

$$u = F(\lambda, v),$$

with u the solution of

$$(\phi(u'))' = -\lambda f(t, v),$$
$$u(a) = 0 = u(b).$$

We note that $F(0, v) = 0$, $v \in C^0[a, b]$. Hence it follows from the continuation theorem of Leray-Schauder that there exists a solution continuum $\mathcal{C} \subset [0, \infty) \times C^0[a, b]$ of solutions of (83) which is unbounded in $[0, \infty) \times C^0[a, b]$, and thus, (83) has a solution for $\lambda > 0$ sufficiently small, and hence $\lambda^* > 0$. By Theorem 6.2, $\lambda^* < \infty$. We verify that $\lambda^* \in \Lambda$. Let $\{\lambda_n\}_n \subset \Lambda$ be such that $\lambda_n \to \lambda^*$ and let $\{u_n\}$ be the corresponding solutions. We easily see that $\{u_n\}$ is bounded in $C^1[a, b]$ and hence $\{u_n\}$ has a subsequence converging to $u \in C^0[a, b]$. By standard limiting procedures, it follows that u is a solution of (83).

Lemma 6.5 *Let $0 < \lambda < \lambda^*$ and let u_{λ^*} be a solution of (83). Then there exists $\epsilon_0 > 0$ such that $u_{\lambda^*} + \epsilon$, $0 \leq \epsilon \leq \epsilon_0$ is an upper solution of (83).*

We now employ the above results to proof Theorem 6.1.

Let $0 < \lambda < \lambda^*$. Since 0 is a lower solution and u_{λ^*} is an upper solution, there exists a minimum solution u_λ of (83) with $0 \leq u_\lambda \leq u_{\lambda^*}$. We next establish the existence of a second solution.

We remark that the mapping

$$\lambda \mapsto u_\lambda, \ 0 \leq \lambda \leq \lambda^*,$$

where u_λ is the minimal solution of $(1.1)_\lambda$ is a continuous mapping $[0, \lambda^*] \to C^0[a, b]$. Hence

$$\{(\lambda, u_\lambda) : \ 0 \leq \lambda \leq \lambda^*\} \subset \mathcal{C},$$

where $\mathcal{C}$ is the continuum in the proof of Lemma 6.4. Using separation results on closed sets in compact metric spaces (Whyburn's lemma), one may use the arguments used in the above proof to verify that for each $\lambda \in (0, \lambda^*)$ there are at least two solutions on the continuum $\mathcal{C}$.

7   On principal eigenvalues for general domains

In this section we shall briefly consider the eigenvalue problem

$$-\text{div}(A(|\nabla u|^2)\nabla u) = \lambda B(|u|^2)u \quad \text{in } \Omega$$
$$u = 0 \quad \text{on } \partial\Omega, \tag{86}$$

where Ω is a bounded smooth domain in $\mathbb{R}^N$ and the functions $\phi(s) = A(|s|^2)s$, $\psi(s) = B(|s|^2)s$ are increasing homeomorphisms which satisfy the following growth conditions:

$$c|s|^{p-1} \leq \phi(s), \psi(s) \leq d|s|^{p-1}, \tag{87}$$

where c, d are positive constants and, as before $p > 1$. We are interested in the existence of values of λ such that (86) has nontrivial positive and negative solutions.

To establish the existence of such eigenvalues, we minimize an appropriate functional subject to a constraint in the Sobolev space $W_0^{1,p}(\Omega)$. We establish the following theorem.

Theorem 7.1 *Let ϕ and ψ be monotone increasing functions which satisfy (87). Then for every constant $\mu > 0$ there exists $\lambda > 0$ such that (86) has a positive (negative) solution u and $\|u\|_{L^p} = \mu$. The set of such λ is bounded away from 0 and bounded above.*

Proof Define the following functions

$$\alpha(s) = \frac{1}{2}\int_0^{s^2} A(s)ds, \quad \beta(s) = \frac{1}{2}\int_0^{s^2} B(s)ds, \tag{88}$$

and the functionals

$$f(u) = \int_\Omega \alpha(|\nabla u|)dx, \quad g(u) = \int_\Omega \beta(|u|)dx. \tag{89}$$

These functional are well defined on $W_0^{1,p}(\Omega)$ because of the growth conditions (87). Furthermore, since f and g are convex and continuous they are weakly lower semi continuous. Since the embedding $W_0^{1,p}(\Omega) \hookrightarrow L^p(\Omega)$ is compact we have that for every constant d the sets

$$S_\mu = \{u \in W_0^{1,p}(\Omega) : g(u) = \mu\},$$

are weakly closed. Also the functional f is coercive on $W_0^{1,p}(\Omega)$, hence the constraint minimization problem

$$f(u) = \min_{v \in S_\mu} f(v) \tag{90}$$

has a solution u (see e.g. [62]). Since both f and g are C^1 functionals, Liusternik's theorem on Lagrange multipliers (see again [62]) implies that there exists λ such that

$$f'(u) + \lambda g'(u) = 0. \tag{91}$$

(Note that $g'(u) \neq 0$, $u \in S_\mu$, $\mu \neq 0$.) On the other hand equation (91) means that

$$-\int_\Omega A(|\nabla u|^2) \nabla u \cdot \nabla v + \lambda \int_\Omega B(u^2) uv = 0, \quad \forall v \in W_0^{1,p}(\Omega), \tag{92}$$

i.e. u is a weak solution of (86). Choosing $u = v$ in (92), we obtain that $\lambda > 0$. Since $f(u) = f(|u|)$ and $g(u) = g(|u|)$, we may assume that u is onesigned in Ω. We next denote by λ_1 the principal eigenvalue of

$$\begin{cases} -\text{div}((|\nabla u|^{p-2})\nabla u) = \lambda |u|^{p-2} u & \text{in } \Omega \\ \\ u = 0 & \text{on } \partial\Omega. \end{cases}$$

Then it is known that λ_1 may be characterized as

$$\lambda_1 = \inf\{\int_\Omega |\nabla u|^p : \int_\Omega |u|^p \neq 0\}, \tag{93}$$

where the constraint equivalently may be replaced by

$$\int_\Omega |u|^p = \kappa,$$

for any $\kappa > 0$.

Let now λ be as above. Then it follows from (92) that

$$\begin{aligned} \lambda &= \frac{\int_\Omega A(|\nabla u|^2)|\nabla u|^2}{\int_\Omega B(u^2) u^2} \\ &\geq \frac{c \int_\Omega |\nabla u|^p}{d \int_\Omega |u|^p} \\ &\geq \frac{c}{d} \lambda_1. \end{aligned} \tag{94}$$

Next we note that, since

$$\lambda \int_\Omega B(u^2) u^2 = \int_\Omega A(|\nabla u|^2)|\nabla u|^2$$

and

$$\beta(|u|) \leq B(u^2) u^2,$$

it follows that

$$\lambda \leq \frac{1}{\mu} \int_\Omega A(|\nabla u|^2)|\nabla u|^2.$$

Also since

$$\frac{c}{p} |u|^p \leq \beta(|u|) \leq \frac{d}{p} |u|^p$$

it follows that there exists a constant $\nu = \nu(\mu)$ such that $\nu\xi \in S_\mu$, where ξ is a fixed eigenfunction corresponding to λ_1, i.e a solution to (93), say, chosen so that $\int_\Omega |\nabla \xi|^p = 1$. Hence

$$\lambda \leq \tfrac{1}{\mu} \int_\Omega A(|\nabla u|^2)|\nabla u|^2$$
$$\leq \tfrac{1}{\mu} \int_\Omega A(\nu^2|\nabla \xi|^2)\nu^2|\nabla \xi|^2$$
$$\leq \tfrac{d}{\mu}\nu^p.$$

We thus have obtained an upper bound for λ.

To obtain higher eigenvalues for equation (86) one may proceed via a Liusternik-Schnirelman approach patterned after the results in [7], [12], [42], [63].

Remark 7.1 *Returning to problem (86) we remark that we can treat equally well the case that ψ satisfies the growth condition*

$$c|s|^{q-1} \leq \psi(s) \leq d|s|^{q-1}, \tag{95}$$

where c, d are positive constants and, as before $q > 1$, with

$$q < \frac{Np}{N-p}.$$

One proceeds ina similar manner noting that the embedding $W_0^{1,p}(\Omega) \hookrightarrow L^q(\Omega)$ is compact and one obtains a result similar to Theorem 7.1 with the exception that the set of principal eiegenvalues no longer is bounded above nor below.

In order to treat more general classes of bifurcation problems including related variational inequalities one may proceed as detailed in [11], [22], [44], [48]. Concerning nonlinear perturbations of *resonant* problems à la Landesman-Lazer, results and procedures are given in [45], [46], [47].

For related results, see also [37], [54] and [56], and for bifurcation problems with the underlying domain $\mathbb{R}^N$ we refer to [2], [23], [24] [25].

8 Bibliography

References

[1] R. ADAMS: *Sobolev Spaces*, Acad. Press, New York, 1975.

[2] W. ALLEGRETTO AND Y. HUANG: *Eigenvalues of the indefinite-weight p-Laplacian in weighted spaces*, preprint, 1995.

[3] A. ANANE: *Simplicité et isolation de la première valeur propre du p-Laplacien*, C. R. Acad. Sci Paris, 305 (1987), pp. 725–728.

[4] A. ANANE: Thèse de doctorat, Université Libre de Bruxelles, 1988.

[5] G. BARLES: *Remarks on uniqueness results of the first eigenvalue of the p-Laplacian*, Ann. Fac. Sci. Toulouse Math., 9 (1988), pp. 65–75.

[6] A. BEN-NAOUM AND C. DE COSTER: *On the existence and multiplicity of positive solutions of the p-Laplacian separated boundary value problems*, preprint, 1995.

[7] P. BINDING AND Y. HUANG: *Two parameter problems for the p-Laplacian*, preprint 1995.

[8] H. BRÉZIS: *Operateurs Maximaux Monotones*, North Holland, 1973.

[9] F. BROWDER: *Nonlinear monotone operators and convex sets in Banach spaces*, Bull. Amer. Math. Soc., 71 (1965), pp. 780–785.

[10] F. BROWDER: *Fixed point theory and nonlinear problems*, Bull. Amer. Math. Soc., 9 (1983), pp. 1–39.

[11] S. CARL: *Leray-Lions operators perturbed by state-dependent subdifferentials*, preprint 1995.

[12] C. COFFMAN: *Lyusternik-Schnirelman theory and eigenvalue problems for monotone potential operators*, J. Funct. Anal. 14(1973), 237-252.

[13] C. DE COSTER: *Pairs of positive solutions for the one dimensional p-Laplacian*, Nonl. Anal., TMA, 23(1994), 669-681.

[14] C. DE COSTER AND P. HABETS: *Upper and lower solutions in the theory of ode boundary value problems: Classical and recent results*, preprint, 1995.

[15] F. DE THÉLIN: *Sur l'espace propre associé à la première valeur propre du pseudo-Laplacien*, C. R. Acad. Sci. Paris, 303(1986), pp. 355–358.

[16] H. DANG AND K. SCHMITT: *Existence of positive solutions for semilinear elliptic equations in annular domains*, Diff. Integral Equations, 7(1994), 747-758.

[17] H. DANG, R. MANÁSEVICH AND K. SCHMITT: *Positive radial solutions of some nonlinear partial differential equations*, Math. Nachr., to appear.

[18] H. DANG, K. SCHMITT, AND R. SHIVAJI: *On the number of solutions of boundary value problems involving the p-Laplacian and similar nonlinear operators*, Electr. J. Diff. Eq., to appear.

[19] K. DEIMLING: *Nonlinear Functional Analysis*, Springer, Berlin, 1985.

[20] M. DEL PINO, M. ELGUETA, AND R. F. MANÁSEVICH: *A homotopic deformation along p of a Leray-Schauder degree result and existence for $(|u'|^{p-2}u')' + f(t,u) = 0$, $u(0) = u(T) = 0$, $p > 1$*, J. Diff. Equa., 80 (1989), pp. 1–13.

[21] M. DEL PINO AND R. F. MANÁSEVICH: *Global bifurcation from the eigenvalues of the p-Laplacian*, J. Diff. Equa., 92 (1991), pp. 226–251.

[22] P. DRÁBEK: *Solvability and Bifurcations of Nonlinear Equations*, Longman Scientific Series, Essex, 1992.

[23] P. DRÁBEK AND Y. HUANG: *Perturbed p-Laplacian in* $\mathbb{R}^N$ *: Bifurcation from the principal eigenvalue*, preprint, 1995.

[24] P. DRÁBEK AND Y. HUANG: *Multiple positive solutions of quasilinear elliptic equations in* $\mathbb{R}^N$, preprint, 1995.

[25] P. DRÁBEK AND Y. HUANG: *Bifurcation problems for the p-Laplacian in* $\mathbb{R}^N$, preprint, 1995.

[26] G. DUVAUT AND J. L. LIONS: *Les Inéquations en Mécanique et en Physique*, Dunod, Paris, 1972.

[27] L. ERBE AND K. SCHMITT: *On radial solutions of some semilinear elliptic equations*, Diff. Integral Equations, 1(1988), 71-78.

[28] A. FRIEDMAN: *Variational Principles and Free Boundary Value Problems*, Wiley-Interscience, New York, 1983.

[29] N. FUKAGAI, M. ITO, AND K. NARUKAWA: *Bifurcation of radially symmetric solutions of degenerate quasilinear elliptic equations*, Diff. Int. Eq. 8(1995), 1709-1732.

[30] M. GARCÍA-HUIDOBRO, R. MANÁSEVICH AND K. SCHMITT: *Positive radial solutions of nonlinear elliptic-like partial differential equations on a ball*, preprint, 1995.

[31] M. GARCÍA-HUIDOBRO, R. MANÁSEVICH AND K. SCHMITT: *On the principal eigenvalue of p-Laplacian like operators*, preprint, 1995.

[32] M. GARCÍA-HUIDOBRO, R. MANÁSEVICH AND K. SCHMITT: *Some bifurcation results for a class of p-Laplacian like operators*, preprint, 1995.

[33] M. GARCÍA-HUIDOBRO, R. MANÁSEVICH AND P. UBILLA: *Existence of positive solutions for some Dirichlet problems with an asymptotically homogeneous operator*, Electronic J. Diff. Eq., 10(1995), 1-22.

[34] M. GARCÍA-HUIDOBRO, R. MANÁSEVICH AND F. ZANOLIN: *A Fredholm like result for strongly nonlinear ode's*, J. Diff. Equations, 114(1993), 132-167.

[35] M. GARCÍA-HUIDOBRO AND P. UBILLA: *Multiplicity of solutions for a class of nonlinear 2nd order equations*. Nonlinear Analysis, TMA, to appear.

[36] D. GILBARG AND N. TRUDINGER: *Elliptic Partial Differential Equations of Second Order*, Sringer, Berlin, 1983.

[37] Z. GUO: *Existence and uniqueness of positive radial solutions for a class of quasilinear elliptic equations*, Appl. Anal., 47(1992), 173-189.

[38] G. GUSTAFSON AND K. SCHMITT: *Nonzero solutions of boundary value problems for second order ordinary and delay differential equations*, J. Diff. Equations, 12(1972), 125-147.

[39] J. HEINONEN, T. KILPELÄINEN AND O. MARTIO: *Nonlinear Potential Theory for Degenerate Elliptic Equations*, Cambridge Univ. Press, Cambridge, 1993.

[40] H. KAPER, M. KNAAP, AND M. KWONG: *Existence theorems for second order boundary value problems*, Diff. Integral Equations, 4(1991), 543-554.

[41] D. KINDERLEHRER AND G. STAMPACCHIA: *An Introduction to Variational Inequalities*, Academic Press, New York, 1980.

[42] M. KRASNOSELS'KII: *Topological Methods in the Theory of Nonlinear Integral Equations*, Pergamon Press, Oxford, 1963.

[43] M. KRASNOSELS'KII: *Positve Solutions of Operator Equations*, Noordhoff, Groningen, 1964.

[44] V. LE: *Some global bifurcation results for variational inequalities*, preprint, 1995.

[45] V. LE AND K. SCHMITT: *Minimization problems for noncoercive functionals subject to constraints*, Trans. Amer. Math. Soc., 347(1995), 4485-4513.

[46] V. LE AND K. SCHMITT: *Minimization problems for noncoercive functionals subject to constraints II*, Advances Diff. Eq., to appear.

[47] V. LE AND K. SCHMITT: *On minimizing noncoercive functionals on weakly closed sets*, Banach Center Publications, to appear.

[48] V. LE AND K. SCHMITT: *On Global Bifurcation for Variational Inequalities*, manuscript, 1995.

[49] J. LIONS: *Quelques Méthodes de Résolution des Problèmes aux Limites non Linéaires*, Dunod, Paris, 1969.

[50] J. LIONS AND E. MAGENES: *Non Homogeneous Boundary Value Problems and Applications*, Springer, Berlin, 1972.

[51] R. MANÁSEVICH AND F. ZANOLIN: *Time mappings and multiplicity of solutions for the one dimensional p-Laplacian*, Nonlinear Analysis, TMA, 21(1993), 269-291.

[52] S. MAIER-PAAPE: *Convergence for radially symmetric solutions of quasilinear elliptic equations is generic,* preprint, 1995.

[53] R. MANÁSEVICH, R. NJOKU AND F. ZANOLIN: *Positive solutions for the one dimensional p-Laplacian,* Diff. Integral Equations, 8(1995), 213-222.

[54] K. NARUKAWA, T. SUZUKI: *Nonlinear eigenvalue problems for a modified capillary surface equation,* Funk. Ekvacioy 37(1994), 81-100.

[55] P. OMARI AND F. ZANOLIN: *Infinitely many solutions of a quasilinear elliptic problem with an oscillatory potential,* preprint, 1995.

[56] M. ÔTANI: *Existence and nonexistence of nontrivial solutions of some nonlinear degenerate elliptic equations,* J. Funct. Anal. 76(1988), 140-159.

[57] D. PASCALI AND J. SBURLAN: *Nonlinear Mappings of Monotone Type,* Sijthoff and Noordhoff, Bucharest, 1978.

[58] P. RABINOWITZ: *Some global results for nonlinear eigenvalue problems,* J. Funct. Anal. 7(1971), 487-513.

[59] P. RABINOWITZ: *Méthodes topologiques et problèmes aux limites nonlinéaires,* Univ. Paris 6 Lab. Analyse Numérique 75010, 1975.

[60] K. SCHMITT: *Boundary value problems for quasilinear second order elliptic boundary value problems,* Nonlinear Analysis, TMA, 2(1978), 263-309.

[61] K. SCHMITT: *Positive solutions of semilinear elliptic equations,* pp. 447-500 in *Toplogical Methods in Nonlinear Differential Equations and Inclusions,* Granas/Frigon editors, Kluwer, Boston, 1995.

[62] M. STRUWE: *Variational Methods,* Springer-Verlag, New York, 1990.

[63] A. SZULKIN: *Ljusternik-Schnirelmann theory on C^1 − manifolds,* Ann. Inst. Henri Poincaré, 5(1988), 119-139.

[64] E. ZEIDLER: *Nonlinear Functional Analysis and its Applications, Vol.I: Fixed-Point Theorems,* Springer, Berlin, 1986.

[65] E. ZEIDLER: *Nonlinear Functional Analysis and its Applications, Vol.IIB: Nonlinear Monotone Operators,* Springer, Berlin, 1990.

BOUNDED SOLUTIONS OF NONLINEAR ORDINARY DIFFERENTIAL EQUATIONS

J. Mawhin

Catholic University of Louvain, Louvain-la-Neuve, Belgium

1 Introduction

A classical result states that if A is a $n \times n$ real matrix and $T > 0$, then the system

$$x'(t) = Ax(t) + p(t), \tag{1}$$

has a T-periodic solution for each T-periodic continuous forcing term p if and only if no eigenvalue of A has the form $ik\omega$ with $k \in \mathbb{Z}$ and $\omega = \frac{2\pi}{T}$. The homogeneous part of equation (1) is then said to be *non-resonant*. In this case, if $h : \mathbb{R} \times \mathbb{R}^n \to \mathbb{R}^n$ is continuous, bounded and T-periodic with respect to t, an easy application of the Schauder fixed point theorem implies that the perturbed system

$$x'(t) = Ax(t) + h(t, x(t)), \tag{2}$$

has at least one T-periodic solution. In this situation of a nonlinear perturbation of a non-resonant linear equation, only some *quantitative* assumption (here the bounded-ness) upon h is necessary to save the existence of a T-periodic solution.

Such a T-periodic solution is of course a special case of a solution which is bounded over $\mathbb{R}$. When p is only assumed to be bounded (or essentially bounded) over $\mathbb{R}$, one can raise the question of the existence of solutions of (1) which are *bounded* over $\mathbb{R}$. A classical result of Perron [26], that we shall recall in Section 2, states that such a solution will exist if no eigenvalue of A lies on the imaginary axis. In this case, it will be shown in Section 3 that the perturbed problem (2) still has a solution bounded over $\mathbb{R}$ for any bounded and continuous nonlinear perturbation h. This can be done using either Schauder-Tikhonov fixed point theorem or Schauder theorem followed by a limiting process. Again, a *quantitative* assumption (boundedness) upon the nonlinear perturbation suffices to guarantee the existence of a bounded solution.

When p is T-periodic and A has eigenvalues of the form $ik\omega$ for some $k \in \mathbb{Z}$, necessary and sufficient conditions upon p are known which insure the existence of a T-periodic solution for (1) *(Fredholm alternative)*. Then, as shown independently by Gaetano Villari [34] and Lazer [17] respectively for third order and second order equations, some *qualitative* assumptions upon h (like sign conditions) have to be added to the quantitative ones (like boundedness) to insure the existence of a T-periodic to (2). For example, in the simplest case of the first order scalar differential equation

$$x'(t) = h(t, x(t)), \tag{3}$$

with, for simplicity, h bounded, a sufficient condition for the existence of a T-periodic solution to (3), when h is T-periodic with respect to t, is that

$$\int_0^T \limsup_{x \to +\infty} h(t, x)\, dt < 0 < \int_0^T \liminf_{x \to -\infty} h(t, x)\, dt,$$

or that

$$\int_0^T \limsup_{x \to -\infty} h(t, x)\, dt < 0 < \int_0^T \liminf_{x \to +\infty} h(t, x)\, dt,$$

(see [19]). Conditions of this type are generally referred as *Landesman-Lazer conditions,* and have been obtained for various classes of nonlinear perturbations of *resonant* linear equations and systems.

The obtention of some Landesman-Lazer conditions for the case of bounded solutions has been initiated by Ahmad [1] in 1991 for some *dissipative* scalar second order differential equations, using some qualitative techniques for the study of those equations, described for example in [29, 27, 39]. Those results have been extended, using a *functional-analytic approach* based upon earlier work of Krasnosel'skii [15, 16], by Ortega, Tineo and Alonso in [24, 25, 2], and by Ward and the author in [20], where upper and lower solutions techniques have been used as well, allowing to deal with some *nondissipative* equations. This functional-analytic approach will be described in Sections 4 to 7 of those lectures, with the hope of stimulating further research in an area where many problems remain unsolved. Other techniques, like Conley index, are also useful to study bounded solutions, but will not be considered here. The reader can consult [35, 36, 37, 38, 21].

If $k \geq 0$, and $n \geq 1$ are integers, we shall denote by BC_n^k (BC^k when $n = 1$ and BC_n when $k = 0$) the space of all continuous mappings $x : \mathbb{R} \to \mathbb{R}^n$ which are of class C^k and are bounded over $\mathbb{R}$, together with all their derivatives of order smaller or equal to k, with the uniform norm

$$\|x\|_k = \sum_{j=0}^{k} \sup_{\mathbb{R}} |x^{(j)}|,$$

or any equivalent one, where $|\cdot|$ denotes some norm in $\mathbb{R}^n$, and $x^{(j)}$ the jth derivative of x. We simply write $\|x\|$ when $k = 0$. It is well known that BC_n^k is a Banach space. We shall also denote by L_n^∞ the space of measurable and essentially bounded (equivalence classes of) mappings from $\mathbb{R}$ to $\mathbb{R}^n$ (L^∞ for $n = 1$), endowed with its classical norm $|\cdot|_\infty$.

Let $f : \mathbb{R} \times \mathbb{R}^n \to \mathbb{R}^n$ be continuous and let

$$x'(t) = f(t, x(t)) \tag{4}$$

be the associated ordinary differential system. A solution $x \in BC_n$ of (4) will be called a *bounded solution* of (4). In the case where f satisfies only the Carathéodory conditions, the definition of a bounded solution is the same, except that it has to be a solution in the Carathéodory sense. If $g : \mathbb{R} \times \mathbb{R}^n \to \mathbb{R}^n$ is continuous and if

$$y^{(k)}(t) = g(t, y(t), y'(t), \ldots, y^{(k-1)}(t)) \tag{5}$$

is the associated ordinary differential equation, then, consistently, a *bounded solution* of (5) will be a solution y of (5) belonging to BC^{k-1}. A similar remark like above holds if g is only supposed to be a Carathéodory function.

I want to acknowledge the CISM for the efficient hospitality during the lectures and Professor Fabio Zanolin for his superb and friendly organisation. Moreover, I am indebted to M. Cherpion and C. De Coster for their careful reading of the manuscript and useful suggestions.

2 Linear systems with bounded forcing term

To see if the existence of bounded solutions is a frequent phenomenon or not, we can first consider a scalar homogeneous linear differential equation with constant coefficient

$$x'(t) = ax(t), \tag{6}$$

where a is a real number. For $a \neq 0$, its solutions are given by $x(t) = ce^{at}$ and $x = 0$ is the only bounded solution of (6). For $a = 0$, the solutions of (6) are the constant functions and all are bounded. In the more general case of a homogeneous linear system with constant coefficients

$$x'(t) = Ax(t), \tag{7}$$

with A a real $(n \times n)$ matrix, whose eigenvalues are denoted by α_j, $(1 \leq j \leq n)$, one can prove in an analogous way that if no α_j lies on the imaginary axis, system (7) has only $x = 0$ as a bounded solution. When some α_j lies on the imaginary axis, (7) has nontrivial bounded solutions which are indeed periodic or quasi-periodic.

In the case of the nonhomogeneous linear system ·

$$x'(t) = Ax(t) + p(t), \tag{8}$$

with a locally integrable input p, we notice first that (8) has *at most* one bounded solution when no eigenvalue α_j of A lies on the imaginary axis. Indeed, if u and v are bounded solutions of (8), then $u - v$ is a bounded solution of (7) and hence is equal to zero. If $\Re\alpha_j < 0$ for all $1 \leq j \leq n$, then

$$|e^{At}| \leq Ke^{-at}, \tag{9}$$

for some $K > 0$, some $a \in]0, \min_{1 \leq j \leq n} |\Re\alpha_j|]$, and all $t \geq 0$. By the variation of constants formula, each solution x of (8) is such that

$$x(t) = e^{A(t-\tau)}x(\tau) + \int_\tau^t e^{A(t-s)}p(s)\,ds. \tag{10}$$

But, if x is a bounded solution of (8), it follows from (9) that

$$\lim_{\tau \to -\infty} e^{A(t-\tau)}x(\tau) = 0.$$

Consequently, letting $\tau \to -\infty$ in (10), we get

$$x(t) = \int_{-\infty}^t e^{A(t-s)}p(s)\,ds. \tag{11}$$

It is now easy to check that (11) is a bounded solution of (8). As, for $t \geq s$,

$$|e^{A(t-s)}p(s)| \leq Ke^{-a(t-s)}|p|_\infty,$$

and $e^{-a(t-\cdot)}$ is integrable over $]-\infty, t]$ for each $t \in \mathbb{R}$, with

$$\int_{-\infty}^t e^{-a(t-s)}\,ds = \frac{1}{a},$$

we see that the integral in the right-hand member of (11) exists and that

$$|x(t)| \leq \frac{K|p|_\infty}{a}, \tag{12}$$

for all $t \in \mathbb{R}$, so that $x \in BC$. A direct verification then shows that (11) is a solution to (8), and hence is its unique bounded solution.

Similarly, if $\Re\alpha_j > 0$ for all $1 \leq j \leq n$, one can verify that

$$x(t) = -\int_t^{+\infty} e^{A(t-s)}p(s)\,ds, \tag{13}$$

is the unique bounded solution of (8) and also satisfies (12).

This existence and uniqueness result can be extended to the case where no eigenvalue α_j of A lies on the imaginary axis *(hyperbolic case)*. This is due to Perron [26].

Lemma 1. *The system (8) has a unique bounded solution for all $p \in BC_n$ if no eigenvalue α_j of A lies on the imaginary axis. This bounded solution satisfies the inequality*

$$|x(t)| \leq \frac{K_1|p|_\infty}{a}, \tag{14}$$

for some $K_1 > 0$ and $a \in]0, \min_{1 \leq j \leq n} |\alpha_j|]$.

Proof. The assumption upon the eigenvalues of A implies (see e.g. [13]) that there exist supplementary projectors P_- and P_+ on $\mathbb{R}^n$ commuting with A and numbers $K > 0$, $a \in]0, \min_{1 \leq j \leq n} \alpha_j]$ such that

$$|e^{At}P_-x| \leq Ke^{-at}|P_-x|, \ t \geq 0,$$

$$|e^{At}P_+x| \leq Ke^{at}|P_+x|, \ t \leq 0.$$

From (10) we then get

$$P_-x(t) = e^{A(t-\tau)}P_-x(\tau) + \int_\tau^t P_-e^{A(t-s)}p(s)\,ds, \tag{15}$$

$$P_+x(t) = e^{A(t-\tau)}P_+x(\tau) + \int_\tau^t P_+e^{A(t-s)}p(s)\,ds. \tag{16}$$

If now x is a bounded solution of (8), then letting $\tau \to -\infty$ in (15), we get

$$P_-x(t) = \int_{-\infty}^t P_-e^{A(t-s)}p(s)\,ds,$$

and letting $\tau \to +\infty$ in (16) we get

$$P_+x(t) = -\int_t^{+\infty} P_+e^{A(t-s)}p(s)\,ds.$$

It is then easy to verify that

$$x(t) = \int_{-\infty}^t P_-e^{A(t-s)}p(s)\,ds - \int_t^{+\infty} P_+e^{A(t-s)}p(s)\,ds \tag{17}$$

is the bounded solution of (8). ∎

In [18, 8, 9], extensions of those results to nonautonomous linear equations

$$x'(t) = A(t)x(t) + p(t)$$

are given in terms of the concept of *exponential dichotomy*.

In the special case of a non-homogeneous scalar linear differential equation

$$L[y](t) = p(t), \tag{18}$$

associated to the linear differential operator with real constant coefficients a_j, $(1 \leq j \leq k-1)$,

$$L[y](t) = y^{(k)}(t) + a_{k-1}y^{(k-1)}(t) + \ldots + a_1 y'(t) + a_0 y(t), \tag{19}$$

the above result implies the existence of a unique bounded solution for all $p \in L^\infty$ if and only if no zero of the associated *characteristic polynomial*

$$L(\lambda) = \lambda^k + a_{k-1}\lambda^{k-1} + \ldots + a_1\lambda + a_0 \tag{20}$$

lies on the imaginary axis. In this case, the estimate (14) takes the form

$$\|y\|_{k-1} \leq \frac{K_1|p|_\infty}{a}. \tag{21}$$

3   Linear equations with continuous forcing term

In this section we shall consider in more detail the case of the existence of bounded solutions for the scalar linear differential equation (18) when $p : \mathbb{R} \to \mathbb{R}$ is only assumed to be continuous, which we write $p \in C$. Let

$$BP = \{x : \mathbb{R} \to \mathbb{R} : x \text{ is continuous and has a primitive bounded over } \mathbb{R}\}.$$

Notice that $BC \not\subset BP$ and $BP \not\subset BC$, as shown respectively by $p(t) = 1$ and by $p(t) = 2t\cos t^2 = (\sin t^2)'$.

The following result is due to Ortega and Tineo [25].

Lemma 2. *Let $p \in C$ be given and assume that $\lambda = 0$ is a simple zero of (20) and that (20) has no other zero on the imaginary axis. Then equation (18) has a bounded solution if and only if $p \in BP$.*

Proof. We first consider the case where $k = 1$. Then, by our assumptions, $L[y] = y'$ and the result is nothing but the definition of BP.

Assume now that $k \geq 2$. By assumption, $a_0 = 0$. We first prove the necessity. Let y be a bounded solution of (18) (so that $y \in BC^{k-1}$) and set

$$P(t) = \int_0^t p(s)\,ds. \tag{22}$$

Integrating both members of (18), we get

$$y^{(k-1)}(t) - y^{(k-1)}(0) + \ldots + a_1[y(t) - y(0)] = P(t),$$

and so P is bounded as $y \in BC^{(k-1)}$.

We now prove the sufficiency. Let $p \in BC$ and consider the equation

$$u^{(k-1)}(t) + a_{k-1}u^{(k-2)}(t) + \ldots + a_2 u'(t) + a_1 u(t) = P(t), \tag{23}$$

where P is defined in (22). By assumptions, all the zeros of the characteristic polynomial of (23) have nonzero real parts, and hence, by the results of the previous section, equation (23) has a unique bounded solution U (so that $U \in BC^{k-2}$). From the equation, we see immediately that $U \in BC^{k-1}$, and, as $P \in C^1$, that $U \in C^k$, and hence satisfies the differential equation

$$U^k(t) + a_{k-1}U^{k-1}(t) + \ldots + a_1 U'(t) = p(t).$$

Thus U is a bounded solution of (18). ∎

The following special case was proved by Ortega [24].

Corollary 1. Let $p \in C$ be given and $c \neq 0$. Then the equation

$$y''(t) + cy'(t) = p(t) \tag{24}$$

has a bounded solution if and only if $p \in BP$.

We can now state and prove an extension of the results of the previous section which is due to Ortega and Tineo [25].

Lemma 3. Let $p \in C$ be given and assume that (20) has no zero on the imaginary axis. Then the equation (18) has a bounded solution if and only if $p \in BP + BC$.

Proof. Necessity. Let y be a bounded solution of (18). Letting

$$p^* = y^{(k)} \in BP,$$

and

$$p^{**} = a_{k-1}y^{(k-1)} + \ldots + a_1 y' + a_0 y \in BC,$$

we see that $p = p^* + p^{**} \in BP + BC$.

Sufficiency. Assume that $p = p^* + p^{**}$ with $p^* \in BP$ and $p^{**} \in BC$. Let

$$M = \frac{d^k}{dt^k} + b_{k-1}\frac{d^{k-1}}{dt^{k-1}} + \ldots + b_1 \frac{d}{dt}$$

be a linear differential operator of order k satisfying the conditions of Lemma 2. Then the equation

$$M[y](t) = p^*(t)$$

has a bounded solution u, so that $u \in BC^{k-1}$. If we make the change of variable

$$y = z + u,$$

then equation (18) becomes

$$L[z](t) = M[u](t) - L[u](t) + p^{**}(t). \tag{25}$$

Notice that

$$M[u] - L[u] = (b_{k-1} - a_{k-1})u^{(k-1)} + \ldots + (b_1 - a_1)u' - a_0 u \in BC,$$

and hence $M[u] - L[u] + p^{**} \in BC$. Consequently, by the results of the previous section, the equation (25) has a unique bounded solution z, and $z + u$ is therefore a bounded solution of (18). ∎

Notice that, as noticed at the beginning of Section 2, the bounded solution in Lemma 3 is necessary unique.

Corollary 2. *Let $p \in C$ be given and let $b > 0$ and $c \neq 0$ or $b < 0$. Then the equation*

$$y''(t) + cy'(t) + by(t) = p(t)$$

has a bounded solution if and only if $p \in BP + BC$.

Proof. $z = iu$ with u real is a zero of the corresponding characteristic polynomial $\lambda^2 + c\lambda + b$ if and only if $b = u^2$ and $cu = 0$. ∎

In what follows, L will be called *nonresonant* if its characteristic polynomial has no root on the imaginary axis and *resonant* if some root lies on the imaginary axis.

4 Bounded nonlinear perturbations of a nonresonant linear equation

The following result, due to M.A. Krasnosel'skii [15], is an useful tool to prove the existence of bounded solutions of nonlinear differential systems. Let $f : \mathbb{R} \times \mathbb{R}^n \to \mathbb{R}^n$ be continuous and consider the associated differential system (4).

Lemma 4. *Let $(t_n)_{n \in \mathbb{N}}$ be an increasing sequence of positive numbers and $(x_n)_{n \in \mathbb{N}}$ be a sequence of functions from $\mathbb{R}$ to $\mathbb{R}^n$ with the following properties.*
1. *$t_n \to +\infty$ as $n \to \infty$.*
2. *For each $n \in \mathbb{N}$, x_n is a solution of (4) defined on $[-t_n, t_n]$.*
3. *There exists a bounded set $B \subset \mathbb{R}^n$ such that, for each $n \in \mathbb{N}$ and every $t \in [-t_n, t_n]$, one has $x_n(t) \in B$.*
Then system (4) has at least one solution x defined on $\mathbb{R}$ and such that $x(t) \in \overline{B}$ for all $t \in \mathbb{R}$.

Proof. For each $n \in \mathbb{N}$, define $M(n) = \max_{|t| \leq t_n, x \in \overline{B}} |f(t, x)|$. From the equation

$$x_n(t) = x_n(0) + \int_0^t f(s, x_n(s)) \, ds, \ t \in [-t_n, t_n],$$

it follows that

$$|x_n(t') - x_n(t'')| \leq M(n)|t' - t''|$$

for all $t', t'' \in [-t_n, t_n]$. Consequently, $(x_n)_{n \in \mathbb{N}}$ restricted to $[-t_0, t_0]$ is a uniformly bounded and equicontinuous sequence. By Ascoli-Arzelà theorem, it has a subsequence $(x_n^0)_{n \in \mathbb{N}}$ which converges uniformly on $[-t_0, t_0]$. Similarly, the sequence $(x_n^0)_{n \geq 1}$ restricted to $[-t_1, t_1]$ is a uniformly bounded and equicontinuous sequence and has a subsequence $(x_n^1)_{n \geq 1}$ which converges uniformly on $[-t_1, t_1]$. Continuing in this way, we construct a sequence of sequences $(x_n^k)_{n \geq k}$ in such a way that $(x_n^k)_{n \geq k}$ is a subsequence of $(x_n^{k-1})_{n \geq k-1}$ and converges uniformly on $[-t_k, t_k]$. The diagonal sequence $(x_n^n)_{n \in \mathbb{N}}$

converges uniformy on every compact interval to a function x^* defined on $\mathbb{R}$ and such that $x^*(t) \in \overline{B}$ for each $t \in \mathbb{R}$. If $t \in \mathbb{R}$, then, for all n such that $t_n \geq |t|$, we have the identity

$$x_n^n(t) = x_n^n(0) + \int_0^t f(s, x_n^n(s))\, ds,$$

so that, letting $n \to \infty$, we obtain

$$x^*(t) = x^*(0) + \int_0^t f(s, x^*(s))\, ds,$$

and x^* is a solution of (4).   ∎

We apply Lemma 4 to a nonlinear perturbation of a nonresonant linear differential equation.

Lemma 5. *Consider the scalar differential equation*

$$L[y](t) = h(t, y(t)), \tag{26}$$

with L defined in (19), and assume that the following conditions hold.
1. The characteristic polynomial of L has no root on the imaginary axis.
2. $h : \mathbb{R} \times \mathbb{R} \to \mathbb{R}$ is continuous and bounded.
Then (26) has at least one bounded solution.

Proof. For each integer $n \geq 1$, define $h_n : \mathbb{R} \times \mathbb{R} \to \mathbb{R}$ by

$$h_n(t, y) = h(t, y) \text{ for } t \in [-2n, 2n[\text{ and } y \in \mathbb{R},$$

extended $4n$-periodically to $\mathbb{R} \times \mathbb{R}$. Clearly, each h_n is a Carathéodory function and, if M is such that $|h(t, y)| \leq M$ for all $(t, y) \in \mathbb{R} \times \mathbb{R}$, then $|h_n(t, y)| \leq M$ for all $(t, y) \in \mathbb{R} \times \mathbb{R}$ and all integers $n \geq 1$. By assumption 1, the equation $L[y](t) = 0$ has only the trivial $4n$-periodic solution for each integer $n \geq 1$. If follows then from a direct application of Schauder fixed point theorem in the space $C_{4n}^{k-1}(\mathbb{R}, \mathbb{R})$ of $4n$-periodic functions of class C^{k-1} that the equation

$$L[y](t) = h_n(t, y(t)) \tag{27}$$

has at least one $4n$-periodic solution y_n. Now, such a solution is also a bounded solution of (27) and hence, by (21), there exists constants $K_1 > 0$ and $a > 0$ such that

$$\|y_n\|_k \leq \frac{K_1 |h_n(\cdot, y_n(\cdot))|_\infty}{a} \leq \frac{K_1 M}{a}. \tag{28}$$

Consequently, for each $n \geq 1$, y_n is a solution of (26) defined on $[-n, n]$ and such that the inequality (28) holds. Applying then Lemma 4 to the first order system equivalent to (26), we immediately obtain the existence of a bounded solution to (26).   ∎

We now consider the case of the perturbation of a nonresonant linear equation by a bounded nonlinearity and a continuous (not necessarily bounded!) forcing term. The following result is due to Ortega and Tineo [25].

Theorem 1. *Let $L[y]$ be given by (19), $p \in C$ and $g : \mathbb{R} \to \mathbb{R}$ continuous and bounded be given, and assume that the characteristic polynomial of L has no root on the imaginary axis. Then the equation*

$$L[y](t) + g(y(t)) = p(t) \tag{29}$$

has a bounded solution if and only if $p \in BP + BC$.

Proof. Necessity. If y is a bounded solution of (29), then letting $p^* = y^{(k)}$, $p^{**} = a_{k-1}y^{(k-1)} + \ldots + a_1 y' + a_0 y + g(y)$, we see that $p^* \in BP$, $p^{**} \in BC$ and $p = p^* + p^{**}$.

Sufficiency. Let $p = p^* + p^{**}$ with $p^* \in BP$ and $p^{**} \in BC$, and, using Lemma 3, let u be a bounded solution of

$$L[y](t) = p(t).$$

Setting $y = z + u$, our problem is reduced to finding a bounded solution of

$$L[z](t) + g(u(t) + z(t)) = 0,$$

and the existence of such a solution follows from Lemma 5.   ∎

Corollary 3. *Consider the equation*

$$y''(t) + cy'(t) + by(t) + g(y(t)) = p(t), \tag{30}$$

where $p \in C$ and $g : \mathbb{R} \to \mathbb{R}$ is continuous and bounded. If $b > 0$ and $c \neq 0$ or $b < 0$, then (30) has a bounded solution if and only if $p \in BP + BC$.

Proof. This follows immediately from Theorem 1 and the argument of the proof of Corollary 2.   ∎

One can now raise the question of the existence of a bounded solution for (30) when $b = 0$, i.e. for a nonlinear perturbation of a resonant linear differential operator. The study of this more delicate case requires a study of the functions belonging to $BP + BC$.

5   Mean values and integrals of some continuous functions

The results of the previous sections show the interest of the subspace $BP + BC$ of the space C of continuous real functions over $\mathbb{R}$. We shall devote this section to some of its properties.

If $C_T \subset BC$ denotes the space of continuous and T-periodic real functions, then $C_T \not\subset BP$, as shown by the example of a nonzero constant function. But it is well known that $p \in C_T$ has a bounded primitive if and only if its *mean value* or *average* $\bar{p}$ defined by

$$\bar{p} := \frac{1}{T} \int_0^T p(s)\, ds,$$

is equal to zero. Therefore, every $p \in C_T$ can be written as

$$p(t) = (p(t) - \bar{p}) + \bar{p} = p^*(t) + p^{**}, \tag{31}$$

with $p^* \in BP$ and $p^{**} \in BC$. Such a decomposition of an element of C_T into the sum a function with bounded (T-periodic) primitive and of a constant (its mean value) plays an important role in the study of periodic solutions of ordinary differential equations (see e.g. [19]). Notice moreover that, for $p \in C_T$,

$$\bar{p} = \frac{1}{T} \int_t^{t+T} p(s)\, ds$$

for every $t \in \mathbb{R}$.

One can therefore be tempted to extend such a decomposition to the space $A \subset BC$ of bounded continuous functions such that the limit

$$\bar{p} = \lim_{T \to +\infty} \left[\frac{1}{T} \int_t^{t+T} p(s)\, ds \right] \tag{32}$$

exists uniformly in $t \in \mathbb{R}$. Notice that elementary considerations show that this existence is equivalent to the equality

$$\lim_{r \to +\infty} \inf_{t-s \geq r} \frac{1}{t-s} \int_s^t p(u)\, du = \lim_{r \to +\infty} \sup_{t-s \geq r} \frac{1}{t-s} \int_s^t p(u)\, du, \tag{33}$$

together with the existence of both limits.

It is well known (see e.g. [12]) that A contains in particular the space AP of almost periodic functions in the sense of Bohr, i.e. the closure, under the uniform norm on $\mathbb{R}$ of the set of trigonometric polynomials $\sum_{k=1}^{n} [a_k \cos \lambda_k t + b_k \sin \lambda_k t]$. But it is also well known that one can find some $p \in AP$ such that $\bar{p} = 0$ and $p \notin BP$ (see e.g. [12] p. 31). Consequently, for $p \in AP$, the decomposition (31) with $\bar{p}$ given by (32) does not in general provide a splitting into a sum of an element of BP and a constant function. This is indeed a *small divisors* problem due to the fact that, for the generalized Fourier series

$$\bar{p} + \sum_{k=1}^{\infty} (a_k \cos \lambda_k t + b_k \sin \lambda_k t)$$

associated to p, 0 can be an accumulation point of the set of the λ_k.

We shall see however that such a splitting can be done in an 'approximative' and suitable way, for the class $BP + BC$, and hence in particular for A and AP. Following

Ortega and Tineo [25], and motivated by (33), let us associate to each $p \in BP + BC$ its *lower mean value* or *average* $\hat{p}$ defined by

$$\hat{p} = \lim_{r \to +\infty} \inf_{t-s \geq r} \left[\frac{1}{t-s} \int_s^t p(u) \, du \right], \tag{34}$$

and its *upper mean value* or *average* $\tilde{p}$ defined by

$$\tilde{p} = \lim_{r \to +\infty} \sup_{t-s \geq r} \left[\frac{1}{t-s} \int_s^t p(u) \, du \right]. \tag{35}$$

Elementary considerations show that if $p = p^* + p^{**}$ is any decomposition of $p \in BP + BC$ with $p^* \in BP$ and $p^{**} \in BC$, then one has

$$\inf p^{**} \leq \widehat{p^{**}} = \hat{p} \leq \tilde{p} = \widetilde{p^{**}} \leq \sup p^{**}. \tag{36}$$

Of course, if $\bar{p}$ exists in the sense of (32), then by (33) we see immediately that $\hat{p} = \bar{p} = \tilde{p}$.

The following result is due to Ortega and Tineo [25].

Lemma 6. *Let $p \in BP + BC$ be a given function and $\alpha < \beta$ real numbers. Then the following statements are equivalent.*
i. $\alpha < \hat{p} \leq \tilde{p} < \beta$.
ii. There exists a decomposition $p = p^* + p^{**}$ with $p^* \in BP$, $p^{**} \in BC$ and

$$\alpha < \inf p^{**} \leq \sup p^{**} < \beta. \tag{37}$$

Proof. If (ii) holds, then, using (36), we immediately obtain (i).

Conversely, assume that (i) holds, write $p = p_1 + p_2$, with $p_1 \in BP$ and $p_2 \in BC$, and let

$$P_i(t) = \int_0^t p_i(u) \, du, \ (i = 1, 2), \ P(t) = P_1(t) + P_2(t).$$

If $t_1, t_2 \in \mathbb{R}$, then

$$P(t_1) - P(t_2) = P_1(t_1) - P_1(t_2) + (t_1 - t_2)p_2(\tau),$$

for some τ between t_1 and t_2. Consequently,

$$|P(t_1) - P(t_2)| \leq b + a|t_1 - t_2|, \tag{38}$$

where $b = 2|P_1|_\infty$ and $a = |p_2|_\infty$. Let $\epsilon > 0$ be such that

$$\alpha < \hat{p} - 2\epsilon < \tilde{p} + 2\epsilon < \beta.$$

Then there exists $r_- > 0$ such that

$$\inf_{t-s \geq r} \frac{1}{t-s} \int_s^t p(u) \, du > \hat{p} - \epsilon > \alpha + \epsilon,$$

whenever $r \geq r_-$ and there exists $r_+ > 0$ such that

$$\sup_{t-s \geq r} \frac{1}{t-s} \int_s^t p(u)\, du < \tilde{p} + \epsilon < \beta - \epsilon,$$

whenever $r \geq r_+$. Hence, if $T = \max\{r_-, r_+\}$ and $r \geq T$, we have

$$\alpha + \epsilon < \inf_{t-s \geq r} \frac{1}{t-s} \int_s^t p(u)\, du \leq \sup_{t-s \geq r} \frac{1}{t-s} \int_s^t p(u)\, du < \beta - \epsilon,$$

so that, for all $t \in \mathbb{R}$, we have

$$\alpha + \epsilon < \frac{1}{T} \int_t^{t+T} p(u)\, du < \beta - \epsilon. \tag{39}$$

Let us define
$$p^{**}(t) = \frac{1}{T} \int_t^{t+T} p(u)\, du, \quad p^*(t) = p(t) - p^{**}(t),$$

so that, clearly, $p^{**} \in BC$ and (37) holds. To prove that $p^* \in BP$, define

$$P^*(t) = P(t) - \frac{1}{T} \int_t^{t+T} P(u)\, du.$$

Then,

$$(P^*)'(t) = p(t) - \frac{1}{T}[P(t+T) - P(t)] = p(t) - \frac{1}{T} \int_t^{t+T} p(u)\, du = p(t) - p^{**}(t) = p^*(t),$$

so that P^* is a primitive of p^*. Now

$$P^*(t) = P(t) - P(\tau)$$

for some $\tau \in \,]t, t+T[$, and hence, using (38),

$$|P^*(t)| \leq b + a|t - \tau| \leq b + aT,$$

for all $t \in \mathbb{R}$, which shows that $p^* \in BP$. ∎

The following special case was obtained by Ortega [24], and shows that a function with mean value zero is arbitrary close to a function with bounded primitive.

Corollary 4. Let $p \in A$ with $\bar{p} = 0$ and $\epsilon > 0$ be given. Then there exists a decomposition $p = p^* + p^{**}$ with $p^* \in BP$, $p^{**} \in BC$ and $|p^{**}|_\infty < \epsilon$.

Proof. By assumption, $-\epsilon < \hat{p} = \bar{p} = \tilde{p} < \epsilon$, and the result follows from Lemma 6. ∎

6 Bounded restoring nonlinear perturbations of dissipative resonant second order equations

To apply the considerations of the above section to the obtention of bounded solutions of some nonlinear perturbations of dissipative resonant second order equations, we need a result, due to Krasnosel'skii and Perov [14] and refined by Krasnosel'skii [15, 16], which constitutes the *method of guiding functions* for the bounded solutions of a differential system. The proof given here is due to Alonso and Ortega [2]. We denote the inner product in $\mathbb{R}^n$ by $(\cdot, \cdot)$, and use the Euclidian norm.

Lemma 7. *Assume that there exist $V \in C^1(\mathbb{R}^n, \mathbb{R})$ and $\rho_0 > 0$ such that the following conditions hold.*
1. $\lim_{|x| \to \infty} V(x) = +\infty$.
2. *For each $t \in \mathbb{R}$ and each $x \in \mathbb{R}^n$ with $|x| \geq \rho_0$, one has*

$$(\nabla V(x), f(t, x)) \leq 0.$$

Then the following is true :
a. *Every solution of (4) defined at $t_0 \in \mathbb{R}$ is defined and bounded in $[t_0, +\infty[$.*
b. *System (4) has at least one bounded solution.*

Proof. We first prove conclusion a, and distinguish two cases. Let $t_0 \in \mathbb{R}$ and x be a solution of (4) defined at t_0. If $|x(t)| > \rho_0$ for all $t \geq t_0$ where the solution is defined, then

$$\frac{d}{dt} V(x(t)) = (\nabla V(x(t)), f(t, x(t))) \leq 0,$$

so that $V(x(t)) \leq V(x(t_0))$ for all $t \geq t_0$ where the solution is defined. Consequently, by assumption 1, $|x(t)|$ remains bounded for those t and his maximal right existence interval is $[t_0, +\infty[$, so that $\limsup_{t \to +\infty} |x(t)| < +\infty$. In the second case, there will exist $\tau \geq t_0$ such that $|x(\tau)| \leq \rho_0$. Define

$$V_0 = \max\{V(x) : |x| \leq \rho_0\}.$$

By assumption 1, there will exist $\rho_1 > \rho_0$ such that $V(x) > V_0$ whenever $|x| \geq \rho_1$. We will prove that $|x(t)| < \rho_1$ for all $t \geq \tau$. If it is not the case, we can find $t_1^* > t_0^* \geq \tau$ such that $|x(t_0^*)| = \rho_0$, $|x(t_1^*)| = \rho_1$ and $\rho_0 < |x(t)| < \rho_1$ for $t \in]t_0^*, t_1^*[$. Then,

$$V_0 < V(x(t_1^*)) \leq V(x(t_0^*)) \leq V_0,$$

a contradiction.

We now prove conclusion b. For each $n \in \mathbb{N}$, let x_n be a solution of (4) such that $x_n(-n) = 0$. By the proof of part a, x_n is defined over $[-n, +\infty[$ and $|x_n(t)| < \rho_1$ for all $t \in [-n, +\infty[$. Applying Lemma 4 to this sequence completes the proof. ∎

We now consider the equation

$$y''(t) + cy'(t) + g(y(t)) = p(t), \tag{40}$$

where $c > 0$, $g : \mathbb{R} \to \mathbb{R}$ is continuous and $p \in C$. Hence the characteristic equation of $L[y] = y'' + cy'$ has the zero eigenvalue on the imaginary axis and no other one. The following result is due to R. Ortega [24], and corresponds to an attractive restoring force g.

Theorem 2. *Assume that g has finite limits $g(-\infty)$ and $g(+\infty)$ when $y \to -\infty$ and $y \to +\infty$ respectively. Then a sufficient condition for the existence of a bounded solution of (40) is that*

$$p \in BP + BC, \ g(-\infty) < \hat{p} \le \tilde{p} < g(+\infty). \tag{41}$$

Moreover, if g is such that

$$g(-\infty) < g(y) < g(+\infty), \tag{42}$$

for all $y \in \mathbb{R}$, then condition (41) is also necessary.

Proof. *Necessity.* If inequality (42) holds and y is a bounded solution of (40), then letting,

$$p^* = y'' + cy', \ p^{**} = g(y),$$

we have $p = p^* + p^{**}$, $p^{**} \in BC$, and it follows from Lemma 2 that $p^* \in BP$. Moreover

$$\inf_{t \in \mathbb{R}} p^{**}(t) = \inf_{t \in \mathbb{R}} g(y(t)) \ge \min_{[-|y|_\infty, +|y|_\infty]} g > g(-\infty),$$

and, similarly, $g(+\infty) > \sup_{t \in \mathbb{R}} p^{**}(t)$. It then follows from Lemma 6 that the inequalities in (41) hold.

Sufficiency. By adding a suitable constant to p in (40), we can always assume, without loss of generality, that

$$g(-\infty) < 0 < g(+\infty). \tag{43}$$

By assumption (41) and Lemma 6, there exists a decomposition $p = p^* + p^{**}$ of p with $p^* \in BP$, $p^{**} \in BC$ and

$$g(-\infty) < \inf p^{**}(t) \le \sup p^{**}(t) < g(+\infty). \tag{44}$$

By Corollary 1, the linear equation

$$y''(t) + cy'(t) = p^*(t),$$

has a bounded solution u. Letting $y = u + z$, we have to find a bounded solution z of the equation

$$z''(t) + cz'(t) + g(u(t) + z(t)) = p^{**}(t),$$

i.e. a bounded solution $x = (x_1, x_2)$ to the equivalent system

$$x_1'(t) = x_2(t), \quad x_2'(t) = -cx_2(t) - g(u(t) + x_1(t)) + p^{**}(t). \tag{45}$$

By (43) and (44), we can find $\mu > 0$ and $\gamma > 0$ such that

$$z[g(u(t) + z) - p^{**}(t)] \geq \mu|z| - \gamma, \tag{46}$$

for all $t \in \mathbb{R}$ and $z \in \mathbb{R}$. Let us define $V \in C^1(\mathbb{R}^2, \mathbb{R})$ by

$$V(x) = V(x_1, x_2) = c^2 x_1^2 + 2cx_1 x_2 + 2x_2^2 = (cx_1 + x_2)^2 + x_2^2.$$

so that

$$\nabla V(x) = (2c^2 x_1 + 2cx_2, 2cx_1 + 4x_2),$$
$$V(x_1, x_2) \to +\infty \text{ if } |(x_1, x_2)| \to \infty,$$

and, if

$$f(t, x) = (x_2, -cx_2 - g(u(t) + x_1) + p^{**}(t)),$$

we get, using (46),

$$(\nabla V(x), f(t, x)) = -2cx_2^2 - 2cx_1[g(u(t) + x_1) - p^{**}(t)] - 4x_2[g(u(t) + x_1) - p^{**}(t)]$$

$$\leq -2cx_2^2 - 2c[\mu|x_1| - \gamma] + 4|x_2|[|g|_\infty + |p^{**}|_\infty].$$

It is not difficult to show the existence of $\rho_0 > 0$ such that the right-hand member of this inequality will be strictly negative when $|(x_1, x_2)| \geq \rho_0$, and hence the result follows from Lemma 7.   ∎

The following special case was already obtained by Ahmad [1], using another argument.

Corollary 5. *Assume that g has (finite) limits $g(-\infty)$ and $g(+\infty)$ when $y \to -\infty$ and $y \to +\infty$ respectively and that $p \in A$. Then a sufficient condition for the existence of a bounded solution of (40) is that*

$$g(-\infty) < \bar{p} < g(+\infty). \tag{47}$$

Moreover, if g is such that (42) holds, then condition (47) is also necessary.

Example 1. The differential equation

$$y''(t) + cy'(t) + \frac{by(t)}{1 + |y(t)|} = p(t), \tag{48}$$

with $p \in C$, $b > 0$ and $c > 0$ has a bounded solution if and only if $p \in BP + BC$ and

$$-b < \hat{p} \leq \tilde{p} < b.$$

Theorem 2 does not hold for the equation

$$y''(t) + cy'(t) + \frac{by(t)}{1 + |y(t)|^a} = p(t),$$

with $a > 1$, because in this case the nonlinear terms has equal limits at $-\infty$ and $+\infty$. The following existence theorem, which can be found in [20], can handle this type of equation.

Let us consider the equation

$$y''(t) + cy'(t) + h(y(t)) = p(t), \tag{49}$$

where $h : \mathbb{R} \to \mathbb{R}$ is continuous and bounded, $c > 0$ and $p \in BP$. By Lemma 2, equation

$$y''(t) + cy'(t) = p(t)$$

has at least one bounded solution, call it $u(t)$.

Theorem 3. *Assume that there exists $r_0 > 0$, $\sigma \geq 0, \epsilon > 0, M_1 > 0$ and $M_2 > 0$ such that, when $t \in \mathbb{R}$ and $|y| \geq r_0$, one has*

$$yh(y) \geq 0, \tag{50}$$

$$|h(y) - h(y + u(t))| \leq \frac{M_1}{|y|^{\sigma+1+\epsilon}}, \tag{51}$$

and

$$yh(y + u(t)) \geq \frac{M_2}{|y|^\sigma}. \tag{52}$$

Then equation (49) has at least one bounded solution.

Proof. Letting

$$y(t) = u(t) + z(t),$$

and noticing that $z \in BC^1$ if and only if $y \in BC^1$, our problem is equivalent to proving the existence of a bounded solution y for the equation

$$z''(t) + cz'(t) + h(u(t) + z(t)) = 0,$$

i.e. the existence of a solution $(x_1, x_2) \in BC \times BC$ for the equivalent system

$$x_1'(t) = x_2(t), \quad x_2'(t) = -cx_2(t) - h(u(t) + x_1(t)). \tag{53}$$

Let

$$f(t, x_1, x_2) = (x_2, -cx_2 - h(u(t) + x_1)).$$

Take

$$0 < B < \min\{2c, 2c^2\}, \tag{54}$$

and

$$V(x_1, x_2) = \frac{Bc}{2}x_1^2 + Bx_1x_2 + cx_2^2 + 2cH(x_1),$$

where

$$H(y) = \int_0^y h(s)\, ds.$$

Because of (50), we have

$$H(y) \geq M_2,$$

for some $M_2 \in \mathbb{R}$ and all $y \in \mathbb{R}$. Because of (54), $\frac{Bc}{2}x_1^2 + Bx_1x_2 + cx_2^2$ is positive definite, and hence

$$V(x_1, x_2) \to +\infty \quad \text{when } x_1^2 + x_2^2 \to \infty.$$

Now,

$$(\nabla V(x_1, x_2), f(t, x_1, x_2)) = (B - 2c^2)x_2^2 + 2cx_2[h(x_1) - h(u(t) + x_1)] - Bx_1 h(u(t) + x_1)$$

$$\leq (B - 2c^2)x_2^2 + 2cx_2[h(x_1) - h(u(t) + x_1)] + B(r_0 + |u|_\infty)|h|_\infty.$$

Notice that, from (54), we have $B - 2c^2 < 0$. Therefore, if $x_1^2 + x_2^2 \geq 2R^2$, with $R \geq r_0$, and $x_2^2 \geq R^2 \geq r_0^2$, then

$$(B - 2c^2)x_2^2 + 2cx_2[h(x_1) - h(u(t) + x_1)] + B(r_0 + |u|_\infty)|h|_\infty$$

$$\leq (B - 2c^2)x_2^2 + 4c|h|_\infty|x_2| + B(r_0 + |u|_\infty)|h|_\infty < 0,$$

if $R \geq r_1$, for some sufficiently large $r_1 \geq r_0$. If $x_1^2 + x_2^2 \geq 2R^2$, with $R \geq r_0$, and $x_2^2 \leq R^2$, then $x_1^2 \geq R^2 \geq r_0^2$, so that (51) and (52) imply

$$(B - 2c^2)x_2^2 + 2cx_2[h(x_1) - h(u(t) + x_1)] - Bx_1 h(u(t) + x_1)$$

$$\leq (B - 2c^2)x_2^2 + \frac{1}{R^\sigma}\left[2c\frac{M_1}{R^\epsilon} - BM_2\right] < 0,$$

if $R \geq r_2$, for some sufficiently large $r_2 \geq r_0$. Thus, for $R \geq \max\{r_1, r_2\}$, we have

$$(\nabla V(x_1, x_2), f(t, x_1, x_2)) < 0$$

whenever $x_1^2 + x_2^2 \geq 2R^2$, and it suffices to take $\rho_0 = 2^{1/2}R$, to fill the conditions of Lemma 7.   ∎

Example 2. We consider the differential equation

$$y''(t) + cy'(t) + \frac{by(t)}{1 + |y(t)|^a} = p(t),\tag{55}$$

with $c > 0$, $b > 0$, $a \geq 2$ and $p \in BP$. For $y \neq 0$,

$$\left| \frac{y}{1+|y|^a} - \frac{u(t)+y}{1+|u(t)+y|^a} \right|$$

$$= \frac{1}{|y|^{a-1}} \cdot \frac{\left\| 1 + \frac{u(t)}{y} \right|^a - \frac{u(t)}{y|y|^a} - \left(1 + \frac{u(t)}{y} \right) \right|}{\left(1 + \frac{1}{|y|^a} \right)\left(\frac{1}{|y|^a} + \left| 1 + \frac{u(t)}{y} \right|^a \right)},$$

so that

$$b \left| \frac{y}{1+|y|^a} - \frac{u(t)+y}{1+|u(t)+y|^a} \right| \leq \frac{M}{|y|^a}$$

for $|y|$ large. On the other hand,

$$\frac{y(y+u(t))}{1+|y+u(t)|^a} = \frac{1}{|y|^{a-2}} \frac{\left(1 + \frac{u(t)}{y} \right)}{\left(\frac{1}{|y|^a} + \left| 1 + \frac{u(t)}{y} \right|^a \right)},$$

so that

$$b\frac{y(y+u(t))}{1+|y+u(t)|^a} \geq \frac{b}{2|y|^{a-2}},$$

for $|y|$ large. Hence the conditions of Theorem 3 are satisfied with $\sigma = a - 2$ and $\epsilon = 1$, and equation (55) has at least one bounded solution for each $p \in BP$.

Remark 1. By changing t into $-t$, the case where $c < 0$ in Theorems 2 and 3 are reduced to the case where $c > 0$, and the same existence result holds.

Remark 2. The problem of the existence of a bounded solution for (55) when $a \in]1, 2[$ remains open. For $a \in]0, 1[$, one can find some results in [1]. We shall now develop techniques which allow to deal with cases where $b < 0$.

7   Repulsive nonlinear perturbations of some resonant second order equations

The considerations of this section will be based upon a simple proposition related to general existence statements using upper and lower solutions (see e.g. [3, 23, 30, 31, 32]) for the second order differential equation

$$z''(t) + cz'(t) = h(t, z(t)),\tag{56}$$

where $c \in \mathbb{R}$, $h : \mathbb{R} \times \mathbb{R} \to \mathbb{R}$ is continuous and bounded on $\mathbb{R} \times [-r, r]$ for every $r > 0$. We give a direct proof for the sake of completeness.

Lemma 8. *Assume that there exist $r_- < r_+$ such that*

$$h(t, r_-) \leq 0 \leq h(t, r_+),$$

for all $t \in \mathbb{R}$. Then equation (56) has at least one solution $z \in BC^1$ with $r_- \leq z(t) \leq r_+$ for all $t \in \mathbb{R}$.

Proof. For each positive integer n, let us consider the periodic boundary value problem

$$z''(t) + cz'(t) = h(t, z(t)), \quad z(-2n) = z(2n), \quad z'(-2n) = z'(2n). \tag{57}$$

To prove the existence of a solution to this problem, we use the Leray-Schauder continuation theorem (see e.g. [19]) and introduce the homotopy

$$z''(t) + cz'(t) = (1 - \lambda)\left(z(t) - \frac{r_- + r_+}{2}\right) + \lambda h(t, z(t)), \tag{58}$$

$$z(-2n) = z(2n), \quad z'(-2n) = z'(2n), \quad \lambda \in [0, 1[.$$

Let $\Omega = \{z \in C([-2n, 2n]) : r_- < z(t) < r_+ \text{ for all } t \in [-2n, 2n]\}$, and, for some $\lambda \in [0, 1[$, let z be a possible solution of (58) such that $z \in \bar{\Omega}$. We show that then $z \in \Omega$. If not, $r_- \leq z(t) \leq r_+$ for all $t \in [-2n, 2n]$, and $z(t_0) = r_-$ for some $t_0 \in [-2n, 2n]$, or $z(t_1) = r_+$ for some $t_1 \in [-2n, 2n]$. We consider, for definiteness, the first situation, so that $z - r_-$ has a minimum at t_0. If $t_0 \in]-2n, 2n[$, this implies

$$z'(t_0) = 0, \quad z''(t_0) \geq 0,$$

and hence, by (58),

$$0 \leq -(1 - \lambda)\left(\frac{r_+ - r_-}{2}\right) + \lambda h(t_0, r_-) < 0,$$

a contradiction. If $t_0 = -2n$ or $2n$, then $z - r_-$ reaches its minimum at $-2n$ and at $2n$, so that

$$z'(-2n) \geq 0 \geq z'(2n),$$

which, together with the periodicity condition, implies that

$$z'(-2n) = z'(2n) = 0.$$

But then, one must have

$$z''(-2n) \geq 0, \quad z''(2n) \geq 0,$$

and the contradiction occurs like in the previous case. It follows therefore from Leray-Schauder theory that, for each positive integer n, the problem (57) has at least one solution z_n such that

$$r_- \leq z_n(t) \leq r_+,$$

for all $t \in [-2n, 2n]$. Now let us define $h_n : \mathbb{R} \times \mathbb{R} \to \mathbb{R}$ by

$$h_n(t, z) = h(t, z) \text{ whenever } t \in [-2n, 2n[,$$

extended $4n$-periodically for $t \in \mathbb{R}$. Thus h_n is piecewise continuous in t and continuous in z, and hence is a Carathéodory function. If we also extend z_n to $\mathbb{R}$ by $4n$-periodicity, then z_n is a $4n$-periodic solution of the equation

$$z''(t) + cz'(t) = h_n(t, z(t)). \tag{59}$$

Consequently, (z_n, z_n') is a bounded solution of the equivalent system

$$x_1'(t) = x_2(t), \quad x_2'(t) = x_1(t) - cx_2(t) + p_n(t),$$

with $p_n(t) = h_n(t, z_n(t)) - z_n(t)$ piecewise continuous and bounded independently of n. As the eigenvalues of the matrix of the linear part are such that $\lambda_1 < 0 < \lambda_2$, it follows from Lemma 1 that

$$|z_n(t)| + |z_n'(t)| \leq \frac{K_1 |p_n|_\infty}{a},$$

for some positive constants K_1 and a, which do not depend upon n, so that

$$|z_n'(t)| \leq R_1,$$

where R_1 is independent of n. Consequently, for $t \in [-n, n]$, (z_n, z_n') is a solution on $[-n, n]$ of the differential system

$$x_1'(t) = x_2(t), \quad x_2'(t) = -cx_2(t) + h(t, x_1(t)), \tag{60}$$

such that

$$r_- \leq z_n(t) \leq r_+, \quad |z_n'(t)| \leq R_1,$$

for all $t \in [-n, n]$. Lemma 4 then implies that (60) has at least one solution defined on $\mathbb{R}$ such that

$$r_- \leq z(t) \leq r_+, \quad |z'(t)| \leq R_1,$$

for all $t \in \mathbb{R}$.   ∎

We now apply Lemma 8 to the differential equation

$$y''(t) + cy'(t) + g(y(t)) = p(t), \tag{61}$$

where $c \in \mathbb{R}$, $g : \mathbb{R} \to \mathbb{R}$ is continuous and bounded and $p \in C$.

The first result constitutes a nonlinear version of Lemma 3 for equation

$$y''(t) + cy'(t) + by(t) = p(t),$$

when $b < 0$. It is a special case of a result of [20], which extends in various directions earlier ones of Corduneanu [10, 11].

Theorem 4. *Assume that the following conditions hold.*
1. $p \in BP + BC$.
2. *There exist $r_0 > 0$ and $\delta > 0$ such that*

$$yg(y) \le -\delta y^2,$$

whenever $|y| \ge r_0$.
Then equation (61) has at least one bounded solution.

Proof. By Lemma 3, the linear equation

$$y''(t) + cy'(t) - \delta y(t) = p(t),$$

has a bounded solution $u(t)$. Letting

$$y(t) = u(t) + z(t),$$

we see that $y \in BC^1$ if and only if $z \in BC^1$, and y will be a solution of (61) if and only if z is a solution of the equation

$$z''(t) + cz'(t) = \delta u(t) - g(u(t) + z(t)). \tag{62}$$

Now, if we set $r_- = -r_0 - |u|_\infty$, then $u + z \le -r_0$ when $z \le r_-$, and hence

$$-\delta u(t) - g(u(t) + z) \le \delta(u(t) + z) - \delta u(t) = \delta z < 0.$$

Similarly, if we set $r_+ = r_0 + |u|_\infty$, then $u(t) + z \ge r_0$ when $z \ge r_+$, and hence

$$-\delta u(t) - g(u(t) + z) \ge \delta(u(t) + z) - \delta u(t) = \delta z > 0.$$

Thus (62) satisfies the conditions of Lemma 8 and Theorem 4 follows. ∎

Example 3. For each $p \in BP + BC$, $a \in \mathbb{R}$, $b > 0$, $c \in \mathbb{R}$, *Duffing's equation*

$$y''(t) + cy'(t) + ay(t) - by^3(t) = p(t)$$

has at least one bounded solution. This example shows that Theorem 4 generalizes in various directions the existence statement for bounded solutions in [6, 7].

Example 4. For each $p \in BP + BC$, $a > 0$, $b > 0$, $c \in \mathbb{R}$, the *piecewise-linear, asymmetric* or *jumping nonlinearities* equation

$$y''(t) + cy'(t) - ay^+(t) + by^-(t) = p(t)$$

has at least one bounded solution. Here, $y^+ = \max(y, 0)$, $y^- = \max(-y, 0)$.

Assumption 2 of Theorem 4 is of course equivalent to

$$\operatorname{sgn} yg(y) \le -\delta|y|,$$

for $|y| \ge r_0$, and corresponds to the *non-resonance* situation when g is linear. We shall now consider to what extent this condition can be weakened to

$$\operatorname{sgn} yg(y) \le -\delta,$$

for $|y| \ge r_0$, at the expense, of course, of some supplementary conditions upon p. This corresponds to the *resonance* or *Landesman-Lazer* situation, that we shall consider for the differential equation (61) with $c \ne 0$.

Theorem 5. *Assume that the following conditions hold.*
1. $c \neq 0$.
2. $p \in BP + BC$.
3. *There exists $r_0 > 0$ and $\delta_- < \delta_+$ such that*

$$g(y) \geq \delta_+ \text{ for } y \leq -r_0, \ g(y) \leq \delta_- \text{ for } y \geq r_0.$$

4. $\delta_- < \hat{p} \leq \tilde{p} < \delta_+$.
Then equation (61) has at least one bounded solution.

Proof. By assumption 2, assumption 4 and Lemma 6, there exists a decomposition $p = p^* + p^{**}$ with $p^* \in BP$, $p^{**} \in BC$, and

$$\delta_- < \inf p^{**} \leq \sup p^{**} < \delta_+. \tag{63}$$

By assumption 1 and Lemma 2, the linear equation

$$y''(t) + cy'(t) = p^*(t)$$

has a bounded solution $u(t)$. Letting

$$y(t) = u(t) + z(t),$$

we see that $y \in BC^1$ if and only if $z \in BC^1$, and y will be a solution of (61) if and only if z is a solution of the equation

$$z''(t) + cz'(t) = p^{**}(t) - g(u(t) + z(t)). \tag{64}$$

Now, if we set $r_- = -r_0 - \sup u$, then $u(t) + z \leq -r_0$ when $z \leq r_-$, and hence, by assumption 3 and (63),

$$p^{**}(t) - g(u(t) + z) \leq p^{**}(t) - \delta_+ \leq \sup p^{**} - \delta_+ < 0.$$

Similarly, if we set $r_+ = r_0 - \inf u$, then $u(t) + z \geq r_0$ when $z \geq r_+$, and hence, by assumption 3 and (63),

$$p^{**}(t) - g(u(t) + z) \geq p^{**}(t) - \delta_- \geq \inf p^{**} - \delta_- > 0.$$

Thus (64) satisfies the conditions of Lemma 8 and Theorem 5 follows.   ∎

Theorem 5 has the following easy consequence for Duffing's equations.

Corollary 6. *Let $c \neq 0$, $g : \mathbb{R} \to \mathbb{R}$ continuous and $p \in BP + BC$. If*

$$\limsup_{y \to +\infty} g(y) < \hat{p} \leq \tilde{p} < \liminf_{y \to -\infty} g(y), \tag{65}$$

then equation (61) has a bounded solution.

Notice that condition $p \in BP + BC$ is *necessary* for (61) to have a bounded solution, because if y is such a solution, then

$$p(t) = y''(t) + cy'(t) + g(y(t)) = p^*(t) + p^{**}(t),$$

with $p^* = y'' + cy' \in BP$ because of Lemma 2.

Corollary 6 implies in particular that if

$$\liminf_{y \to -\infty} g(y) = +\infty, \quad \limsup_{y \to +\infty} g(y) = -\infty, \tag{66}$$

then (61) has a bounded solution for each $p \in BP + BC$.

If

$$-\infty < \limsup_{y \to +\infty} g(y) < g(y) < \liminf_{y \to -\infty} g(y) < +\infty, \tag{67}$$

for all $y \in \mathbb{R}$, then $p \in BP + BC$ and (65) is also a necessary condition for the existence of a bounded solution of (61). Indeed, if (61) has a bounded solution y, then

$$p(t) = y''(t) + cy'(t) + g(y(t)) = p^*(t) + p^{**}(t),$$

with $p^* = y'' + cy' \in BP$ because of Lemma 2 and $p^{**} = g(y) \in BC$. The result then follows from Lemma 6.

Example 5. Equation

$$y''(t) + cy'(t) - b\frac{y(t)}{1 + |y(t)|} = p(t),$$

with $c \neq 0$ and $b > 0$, has a bounded solution if and only if $p \in BP + BC$ and

$$-b < \hat{p} \leq \tilde{p} < b.$$

Example 6. Equation

$$y''(t) + cy'(t) - b\frac{y(t)}{1 + |y(t)|^a} = p(t),$$

with $c \neq 0$, $b > 0$ and $0 \leq a < 1$ has a bounded solution if and only if $p \in BP + BC$.

Remark 3. Using Theorem 5 applied to each equation

$$y''(t) + \frac{1}{n}y'(t) + g(y(t)) = p(t),$$

followed by a delicate limit process, one can prove (see [20]) the existence of a bounded solution for the equation

$$y''(t) + g(y(t)) = p(t)$$

when $p \in BC$, $g : \mathbb{R} \to \mathbb{R}$ is continuous, bounded and such that

$$g(y) \geq \delta_+ \text{ whenever } y \leq -r_0, \ g(y) \leq \delta_- \text{ whenever } y \geq r_0,$$

for some $\delta_- < \delta_+$ and some $r_0 > 0$. Those conditions are in particular satisfied if $p \in BC$ and

$$-\infty < g(+\infty) < \hat{p} \leq \tilde{p} < g(-\infty) < +\infty.$$

For example the equation

$$y''(t) - \frac{by(t)}{1 + |y(t)|} = p(t),$$

has a bounded solution if $b > 0$, $p \in BC$ and $-b < \hat{p} \leq \tilde{p} < b$. We refer to [20] for the details. Is is an open problem to know if one can replace $p \in BC$ by $p \in BP + BC$ or if one can weaken the boundedness assumption upon g. It would be also interesting to have a variational proof of this result.

References

[1] Ahmad, S., *A nonstandard resonance problem for ordinary differential equations*, Trans. Amer. Math. Soc. **323** (1991), 857-875

[2] Alonso, J.M. and Ortega, R., *Global asymptotic stability of a forced Newtonian system with dissipation*, J. Math. Anal. Appl. 196 (1995), 965-986

[3] Barbalat, I., *Applications du principe topologique de T. Wazewski aux équations différentielles du second ordre*, Ann. Polon. Math. **5** (1958), 303-317

[4] Bebernes, J.W., and Jackson, L.K., *Infinite interval boundary value problems for $y'' = f(x, y)$*, Duke Math. J. **34** (1967), 39-48

[5] Belova, M.M., *Bounded solutions of non-linear differential equations of second order*, Matematičeskii Sbornik (NS) **56** (1962), 469-503

[6] Berger, M.S., and Chen, Y.Y., *Forced quasiperiodic and almost periodic oscillations of nonlinear Duffing equations*, J. Nonlinear Anal. **19** (1992), 249-257

[7] Berger, M.S., and Chen, Y.Y., *Forced quasiperiodic and almost periodic oscillations for nonlinear systems*, J. Nonlinear Anal. **21** (1993), 949-965

[8] Coppel, W.A., *Stability and Asymptotic Behavior of Differential Equations*, Heath, Boston, 1965

[9] Coppel, W.A., *Dichotomies in Stability Theory*, Springer, Berlin, 1978

[10] Corduneanu, C., *Existence theorems on the real axis for solutions of some non-linear second order differential equations* (Roumanian), Bul. Stiint. Acad. Rep. Pop. Roum., Sect. stiint. mat. fiz. **7** (1955), 645-651

[11] Corduneanu, C., *A few problems concerning differential equations of the second order with small parameter* (Roumanian), Bul. Stiint. Acad. Rep. Pop. Roum., Sect. stiint. mat. fiz. **8** (1956), 703-707

[12] Corduneanu, C., *Almost Periodic Functions*, Wiley, New York, 1968. Reprinted, Chelsea, New York, 1989

[13] Hale, J.K., *Ordinary Differential Equations*, Wiley-Interscience, New York, 1969

[14] Krasnosel'skii, M.A., and Perov, A.I., *On a certain principle of existence of bounded, periodic and almost periodic solutions of systems of ordinary differential equations*, (Russian), Dokl. Akad. Nauk SSSR **123** (1958), 235-238

[15] Krasnosel'skii, M.A., *Translation along trajectories of differential equations*, (Russian), Nauka, Moscow, 1966. English translation Amer. Math. Soc., Providence, 1968

[16] Krasnosel'skii, M.A., and Zabreiko, P.P., *Geometrical Methods of Nonlinear Analysis*, (Russian), Nauka, Moscow, 1975. English translation Springer, Berlin, 1984

[17] Lazer, A.C., *On Schauder's fixed point theorem and forced second order nonlinear oscillations*, J. Math. Anal. Appl. **21** (1968), 421-425

[18] Massera, J.L., and Schäffer, J.J., *Linear Differential Equations and Function Spaces*, Academic Press, New York, 1966

[19] Mawhin, J., *Topological Methods in Nonlinear Boundary Value Problems*, CBMS Regional Confer. Nr. 40, Amer. Math. Soc., Providence, 1979

[20] Mawhin, J., and Ward, J.R., *Bounded solutions of some second order nonlinear differential equations*, to appear.

[21] Memory, M.C., and Ward, J.R., *The Conley Index and the method of averaging*, J. Math. Anal. Appl. 158 (1991), 509-518

[22] Mukhamadiev, E., Nazhmiddinov, Kh., and Sadovskii, B.N., *Application of the Schauder-Tokhonov principle to the problem of bounded solutions of differential equations*, Functional Analysis and Applications 6 (1972), 246-247

[23] Opial, Z., *Sur les intégrales bornées de l'équation $u'' = f(t, u, u')$*, Ann. Polon. Math. **4** (1958), 314-324

[24] Ortega, R., *A boundedness result of Landesman-Lazer type*, Differential and Integral Equations **8** (1995), 729-734

[25] Ortega R., and Tineo, A., *Resonance and non-resonance in a problem of boundedness*, Proc. Amer. Math. Soc., to appear

[26] Perron, O., *Die Stabilitätsfrage bei Differentialgleichungen*, Math. Z. **32** (1930), 703-728

[27] Reissig, R., Sansone, G., and Conti, R., *Qualitative Theorie nichtlineare Differentialgleichungen*, Cremonese, Roma, 1963

[28] Rouche, N., and Mawhin, J., *Ordinary Differential Equations. Stability and Periodic Solutions*, (French), Masson, Paris, 1973. English translation Pitman, London, 1980

[29] Sansone, G., and Conti, R., *Nonlinear Differential Equations*, (Italian), Cremonese, Roma, 1956. English translation Pergamon, Oxford, 1964

[30] Schmitt, K., *Bounded solutions of nonlinear second order differential equations*, Duke Math. J. **36** (1969), 237-243

[31] Schrader, K.W., *Boundary value problems for second order ordinary differential equations*, J. Differential Equations **3** (1967), 403-413

[32] Schrader, K.W., *Solutions of second order ordinary differential equations*, J. Differential Equations **4** (1968), 510-518

[33] Tineo, A., *An iterative scheme for the N-competing species problem*, J. Differential Equations **116** (1995), 1-15

[34] Villari, G., *Soluzioni periodiche di una classe di equazioni differenziali del terz'ordine quasi lineari*, Ann. Mat. Pura Appl. **73** (1966), 103-110

[35] Ward, J.R., *Averaging, homotopy, and bounded solutions of ordinary differential equations*, Differential and Integral Equations 3 (1990), 1093-1100

[36] Ward, J.R., *A topological method for bounded solutions of non-autonomous ordinary differential equations*, Trans. Amer. Math. Soc. 333 (1992), 709-720

[37] Ward, J.R., *Homotopy and bounded solutions of ordinary differential equations*, J. Differential Equations 107 (1994), 428-445

[38] Ward, J.R., *Global continuation for bounded solutions of ordinary differential equations*, Topological Methods in Nonlinear Anal. 2 (1993), 75-90

[39] Yoshizawa, T., *Stability Theory and the Existence of Periodic Solutions and Almost Periodic Solutions*, Springer, New York, 1975

HYPERBOLIC STRUCTURES IN ODE'S AND THEIR DISCRETIZATION
WITH AN APPENDIX ON
DIFFERENTIABILITY PROPERTIES OF THE INVERSION OPERATOR

B.M. Garay

University of Technology, Budapest, Hungary

Abstract. We survey some recent results of numerical dynamics centered about hyperbolic structures including structural stability, normally hyperbolic compact invariant manifolds, and topological horseshoes on transversal sections. The presentation itself is followed by a discussion on some of the underlying abstract mathematical theories. The paper ends with results on the inversion operator we have no references for. Proposition b.) stating that "operator $\varphi \to \varphi^{-1}$ as a self-homeomorphism of $\text{Diff}^1(M)$ is nowhere differentiable" seems to be new.

1 Introduction

The present contribution is devoted to the question if qualitative properties of ODE's are preserved under discretization. This question can not be answered by a simple 'yes' or 'no'. Even in very simple systems, large stepsize may easily lead to chaos. Weak stability properties, symplectic structures, attracting manifolds for singular perturbations etc. are preserved only by carefully selected classes of discretization methods. Fortunately – and this is what we deal with in the sequel – hyperbolic configurations are correctly reproduced by any reasonable one-step method, for stepsize sufficiently small. Both for differential equations and diffeomorphisms, hyperbolic configurations are known to persist under small C^1-perturbations. We investigate some discretization aspects of this general principle and consider various global concepts like structural stability, normally hyperbolic invariant manifolds, orbits connecting hyperbolic periodic orbits including the homoclinic case with horseshoe-type chaos on Poincaré sections. We do not discuss bifurcation questions. Results in Sections 1–3 are continuations of those contained in Chapters 4 and 5 of the beautiful survey of A. M. Stuart [56]. (See also his forthcoming monograph with A. R. Humphries [57].) The proofs involve techniques from differential topology. Discretization mappings on manifolds are defined. Estimates in various manifold topologies are performed on coordinate chart representations.

Our interest is mainly theoretical. Practical aspects of computing play absolutely no role in Sections 1–3. Section 4 is of entirely different character. The most practical aspects of computing like cumulative round-off errors and the actual value of the right-hand side (of case $j = 0$) of inequality (2) become extremely important in Section 4.

We consider ordinary differential equations of the form $\dot{x} = f(x)$ where $f : \mathbf{R}^n \to \mathbf{R}^n$ is of class C^{p+k+1}, with all derivatives bounded, and n, p, k are integers, $n \geq 1$, $p \geq 1$, $k \geq 0$, $p + k \geq 2$. The generated solution flow is denoted by Φ. With the differential equation $\dot{x} = f(x)$, we consider also a C^{p+k+1} discretization mapping $\varphi : [0, h_0] \times \mathbf{R}^n \to \mathbf{R}^n$, with all derivatives bounded, where h_0 is a positive constant and, as usual, the C^{p+k+1} property of φ on $[0, h_0] \times \mathbf{R}^n$ is understood as the existence of a C^{p+k+1} extension $\hat{\varphi}$ defined on some open neighbourhood of $[0, h_0] \times \mathbf{R}^n$ in $\mathbf{R} \times \mathbf{R}^n$. We assume that φ is of order p (sometimes we need that $p \geq 2$ or $p \geq 3$) i.e.

$$|\Phi(h, x) - \varphi(h, x)| \leq \text{const} \cdot h^{p+1} \quad \text{for all} \quad h \in [0, h_0], \quad x \in \mathbf{R}^n. \tag{0}$$

Besides these assumptions, the only requirement on φ is the existence of a continuous function $\Delta : [0, h_0] \to \mathbf{R}^+$ such that $\Delta(0) = 0$ and, for all $x \in \mathbf{R}^n$, $\varphi(h, x)$ depends only on the restriction of f to the set $\{z \in \mathbf{R}^n \,||\, z - x| \leq \Delta(h)\}$. Obviously, all conditions above are satisfied if φ comes from a general r-stage explicit or implicit Runge-Kutta method of order p.

The most important properties of φ are as follows. (For details and proofs, see [25].) First of all, for h sufficiently small, as a direct application of the Hadamard-Levy global inverse function theorem, $\varphi(h, \cdot)$ is a diffeomorphism of $\mathbf{R}^n$ onto itself. Consequently, time can be reversed, and also negative iterates of $\varphi(h, \cdot)$ can be defined. The discrete-time dynamical systems $\{\varphi^M(h, \cdot)\}_{M \in \mathbf{Z}}$ and $\{\Phi(Mh, \cdot)\}_{M \in \mathbf{Z}}$ are approximating each other in the sense that

$$|(\Phi(h, x) - \varphi(h, x))_x^{(j)}| \leq \text{const} \cdot h^{\mu(j)+1} \tag{1}$$

and, with const $(\cdot)$ denoting an appropriate continuous real function,

$$|(\Phi(Mh, x) - \varphi^M(h, x))_x^{(j)}| \leq \text{const}\,(|M|h) \cdot h^{\mu(j)} \tag{2}$$

for all $x \in \mathbf{R}^n$, $M \in \mathbf{Z}$ and h sufficiently small where

$$\mu(j) = \min\{p, p + k - j\}, \quad j = 0, 1, \ldots, p + k.$$

In virtue of the Hadamard-Landau norm interpolation inequality, the finite set of inequalities (2) is equivalent to the collection of its special cases $j = 0$, $j = p$, $j = p + k$.

Let $\mathcal{L}$ denote the supremum of the logarithmic norm of $f'(x)$ over $\mathbf{R}^n$ and assume that $\mathcal{L} \geq 0$. (For the $\mathcal{L} < 0$ case see the first paragraph of Subsection 2.2). The main result of [25] provides an upper bound for const $(|M|h)$ in (2). Given $\varepsilon > 0$ arbitrarily, there exists a constant c_ε such that

$$\text{const } (Mh) \leq c_\varepsilon \cdot \exp((\mathcal{L} + \varepsilon)(j + 1)Mh) \tag{3}$$

for all h sufficiently small and $M \in \mathbf{N}$. Besides $\mathcal{L} \geq 0$, the only property of $\mathcal{L}$ we need is that

$$|\Phi_x'(h, x)| \leq 1 + (\mathcal{L} + \varepsilon)h \tag{4}$$

for all $x \in \mathbf{R}^n$ and h sufficiently small. Actually (3) can be replaced by

$$\text{const } (Mh) \leq c_\varepsilon Mh \cdot \exp((\mathcal{L} + \varepsilon)(j + 1)Mh).$$

(The proof of this somewhat stronger inequality requires only a slight modification of the proof of (3). The last inequality before the triple summation in proving Theorem 1.1 of [25] can be strengthened to

$$\sum_{i=1}^{I(s)} Q_i(s) \leq (1 + c_1 h)^{jMh} a_j e^{\gamma_1 jMh}(q + jMh)h^{q + \nu(j)}$$

and consequently, the last inequality of the whole proof goes over into

$$\sum_{q=0}^{j-1} (q + jMh)(Mh)^q \leq 2j^2 Mh(1 + (Mh)^{j-1}).)$$

Our basic references for numerical ODE methods are J. Butcher [9] and E. Hairer, S. P. Norsett and G. Wanner [30]. For dynamical systems in general, we recommend M. C. Irwin [36] and M. Shub [54]. As for the differential topological background, see M. W. Hirsch [34]. For PDE aspects, see J. K. Hale [31].

2   Results

We compare two continuous-time and four discrete-time dynamical systems as indicated by the diagram

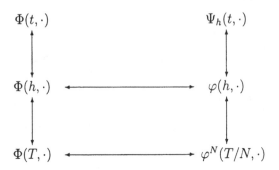

The arrows mean that (for h small, T fixed, N large) the qualitative properties we investigate of the respective dynamical systems are the same. Moreover, in certain cases, discretization is equivalent to the introduction of a new coordinate system: mappings $\Phi(h, \cdot)$ and $\varphi(h, \cdot)$ are connected by a near-to-identity conjugacy $\mathcal{H}_h$ and the conjugacy relation $\mathcal{H}_h(\Phi(h, \cdot)) = \varphi(h, \mathcal{H}_h(\cdot))$ implies that $\varphi(h, \cdot)$ embeds (as its time-h-map) into a new continuous-time dynamical system Ψ_h. We emphasize that "persistence under discretization" does not automatically follow from "persistence under small C^1-perturbations". Standard perturbation theory helps only when comparing $\Phi(T, \cdot)$ and $\varphi^N(T/N \cdot)$. The reason is that, with $h \to 0^+$, both $\Phi(h, \cdot)$ and $\varphi(h, \cdot)$ approach the identity, a degenerate (nonhyperbolic) operator in perturbation theory. What we have to deal with is the one-parameter family of perturbations $\{\varphi(h, \cdot)\}_{h>0}$ of the one-parameter family of diffeomorphisms $\{\Phi(h, \cdot)\}_{h>0}$. Argumentation along vertical arrows of the diagram is usually based on 'weighted' asymptotic phase considerations and/or growth order properties.

2.1. Discretization as a coordinate transformation near nonequilibria. We begin with presenting a discretization analogue of the well-known local flow-box theorem. Fix $x_0 \in \mathbf{R}^n$ with $f(x_0) \neq 0$. Then there is a neighbourhood $\mathcal{N}$ of x_0 and, for h sufficiently small, there exists a C^{p+k+1} diffeomorphism $\mathcal{D}_h : \mathcal{N} \to \mathcal{D}_h(\mathcal{N}) \subset \mathbf{R}^n$ with the properties that $|(\mathcal{D}_h(x) - x)_x^{(j)}| \leq$ const $\cdot h^{\mu(j)}$, $j = 0, 1, \ldots, p + k$ and $\mathcal{D}_h(\Phi(h, x)) = \varphi(h, \mathcal{D}_h(x))$ whenever $x \in \mathcal{N}$, $\Phi(h, x) \in \mathcal{N}$. Moreover, there exists a bounded C^{p+k} function $f_h : \mathbf{R}^n \to \mathbf{R}^n$ such that the solution flow Ψ_h of the differential equation $\dot{x} = f_h(x)$ satisfies $\Psi_h(h, x) = \varphi(h, x)$ for all $x \in \mathcal{N}$ and $|(f_h(x) - f(x))_x^{(j)}| \leq$ const $\cdot h^{\mu(j+1)}$ for all $x \in \mathbf{R}^n$ and $j = 0, 1, \ldots, p + k - 1$. In particular, the approximate solution $\varphi(h, \cdot)$ in $\mathcal{N}$ can be interpreted as (the time-one map of) the exact solution of an ordinary differential equation nearby. This is a justification of the method of modified equations. The differential equation $\dot{x} = f_h(x)$ can be termed as perfectly modified. The proof is based on smooth interpolation (and not as expected on the inverse function theorem). For details, see [27]. We do not know if standard proofs of the local flow-box theorem apply.) All the previous properties hold true if x_0 is replaced by any compact transversal section S of Φ.

Question 1: Assume both Φ and φ are symplectic. Can Ψ_h be chosen to be symplectic, too? An affirmative answer would fit well in the main streamlines of the method of modified equations; see Maria Calvo and J. M. Sanz-Serna [10].

The following result is a somewhat more global version of the very same principle. The proof is a mixture of Liapunov theory, the method of fundamental domains and smooth interpolation techniques. Let A be a nonempty compact asymptotically stable invariant set (briefly: an attractor) of Φ and let $A(M)$ denote its region of attraction. By a basic result of Peter Kloeden and Jens Lorenz [37], the discrete-time dynamical system $\{\varphi^M(h, \cdot)\}_{M \in \mathbf{Z}}$ has such an attractor M_h that, with $h \to 0^+$, M_h approaches M in an upper semicontinuous way. Conditions implying lower semicontinuity are discussed in Chapter 6 of Andrew Stuart [56].

Theorem 1: [27]. For all h sufficiently small, there exists a C^{p+k+1} diffeomorphism $\mathcal{D}_h$ of $A(M)\backslash M$ onto $A(M_h)\backslash M_h$ such that $\mathcal{D}_h(\Phi(h, x)) = \varphi(h, \mathcal{D}_h(x))$ whenever $x \in A(M)\backslash M$. Here of course, $A(M_h)$ denotes the region of attraction of M_h. Further, for any compact subset Q of $A(M)\backslash M$, there exists a constant K_Q such that

$$|(\mathcal{D}_h(x) - x)_x^{(j)}| \leq K_Q h^{\mu(j)} \quad \text{for all} \quad x \in Q \quad \text{and} \quad j = 0, 1, \ldots, p+k \ .$$

It is worth mentioning here that also R_h, the chain-recurrent subset of M_h approaches R, the chain-recurrent subset of M, in an upper semicontinuous way. Lower semicontinuity holds true under the condition that the set of hyperbolic equilibria and hyperbolic periodic orbits in dense in R. (For details and proofs, see our joint paper with Josef Hofbauer [29].)

Question 2: Is lower semicontinuity generic? (The question is open for attractors, too.)

2.2. Discretization as a coordinate transformation near hyperbolic equilibria. If $\mathcal{L} < 0$, then both $\Phi(h, \cdot)$ and $\varphi(h, \cdot)$ define contraction operators on $\mathbf{R}^n$. The unique equilibrium point of $\dot{x} = f(x)$ and the unique fixed point of $\varphi(h, \cdot)$ are denoted by x_0 and x_h, respectively. With $M = \{x_0\}$ and $M_h = \{x_h\}$, the conditions of Theorem 1 are satisfied. By letting $\mathcal{D}_h(x_0) = x_h$, diffeomorphism $\mathcal{D}_h$ extends to a self-homeomorphism of $\mathbf{R}^n = A(x_0) = A(x_h)$ and thus constitutes a conjugacy between $\Phi(h, \cdot)$ and $\varphi(h, \cdot)$ on $\mathbf{R}^n$.

In the vicinity of saddle points, the following version of Grobman-Hartman Lemma holds true. It improves Wolf-Jürgen Beyn's nonconjugacy result [7] on mutual shadowing and generalizes Michal Fečkan's conjugacy result [18] on the Euler method.

Theorem 2: Timo Eirola [17] and (independently) [21]. Fix $x_0 \in \mathbf{R}^n$ with $f(x_0) = 0$ and assume $f'(x_0)$ has no eigenvalues on the imaginary axis. Then there is a neighbourhood $\mathcal{N}$ of x_0 and, for h sufficiently small, there exists a homeomorphism $\mathcal{H}_h : \mathcal{N} \to \mathcal{H}(\mathcal{N}) \subset \mathbf{R}^n$ with the properties that $|\mathcal{H}_h(x) - x| \leq$ const $\cdot h^p$ and $\mathcal{H}_h(\Phi(h,x)) = \varphi(h,\mathcal{H}_h(x))$ whenever $x \in \mathcal{N}$, $\Phi(h,x) \in \mathcal{N}$.

Moreover, the global unstable manifold $W_h^u(x_h)$ of x_h is an injectively C^{p+k+1} immersed copy of the global unstable manifold $W^u(x_0)$ of x_0 and, with $h \to 0^+$, $W_h^u(x_h)$ approaches $W^u(x\ )$ in the C^{p+k-1} weak topology on $C^{p+k+1}(W^u(x_0), \mathbf{R}^n)$. Providing $W^u(x_0)$ is of compact closure, it follows that the limiting process is also lower semicontinuous. The same results hold true for the corresponding global unstable manifolds as well. (For details and proofs, see [25].)

Question 3: In the vicinity of nonhyperbolic equilibria, can one prove discretization results of Shoshitaishvili type? (The general approach of Andrejs Reinfelds [50] to partial linearization as well as the discretization results [8], [21] on center manifolds suggest the answer is affirmative.)

Inequality (2) and all results in this Subsection can readily be generalized for nonautonomous ordinary differential equations. Hyperbolicity has to be replaced by exponential dichotomy and the single conjugacy relation goes over into a doubly-infinite chain of commutative diagrams. (For details, see our joint paper with Bernd Aulbach [5].)

2.3. Structural stability under discretization. The next object to study is a hyperbolic periodic orbit. By a basic result of Wolf-Jürgen Beyn [6] and (somewhat later, in a slightly more general form) Timo Eirola [16], a hyperbolic periodic orbit Γ goes over into a $\varphi(h,\cdot)$-invariant simple closed curve Γ_h (called Beyn-Eirola curve in the sequel) nearby. Since the respective dynamics on Γ and Γ_h can be completely different – rational versus irrational rotation numbers –, the global dynamics of $\Phi(h,\cdot)$ and $\varphi(h,\cdot)$ can not be compared but only on the basis of rather weak equivalence concepts. Among conjugacy results, the following Theorem can not be far from the best possible.

Theorem 3: [26]. Assume $p \geq 2$ and $p+k \geq 3$. Let $\Omega = \{x \in \mathbf{R}^n | |x| < 1\}$ and assume that Φ is Morse-Smale without (nontrivial) periodic orbits in Ω and inward on $\partial\Omega$. Then, for h sufficiently small, there exists a homeomorphism $\mathcal{J}_h : \Omega \to \mathcal{J}_h(\Omega) \subset \mathbf{R}^n$ with the properties that $|\mathcal{J}_h(x) - x| \leq$ const $\cdot h^p$ and $\mathcal{J}_h(\Phi(h,x)) = \varphi(h,\mathcal{J}_h(x))$ whenever $x \in \Omega$.

The natural way of proving Theorem 3 is to lift (via inverse stereographic projection) function f to a vector field on $\mathbf{S}^{n+1}$. This transforms the problem in the traditional framework of structural stability theory. What is actually proven is that given a compact C^∞ manifold M, the triple (ρ, Φ, φ) of a Riemannian metric, a Morse-Smale dynamical system without periodic orbits and a discretization mapping φ of order p on M, there exists a conjugacy $\mathcal{J}_h$ between $\Phi(h,\cdot)$ and $\varphi(h,\cdot)$ with $\max \{\rho(\mathcal{J}_h(x),x) | x \in M\} \leq$ const $\cdot h^p$, h sufficiently small. The proof itself is a reconsideration of the Moser-Robbin-Robinson approach [45], [51], [52] to absolute structural stability, with the necessary estimates supplemented.

The two-dimensional case of Theorem 3 (assuming only that $p \geq 2$) can be proved [23] by standard methods of plane topology. As it is discussed in [23], an affirmative answer to the following question (referring to the possibility of a *quantitative* extension of the classical Schönfliess Theorem, one of the most outstanding *qualitative* results of plane topology) would make this proof considerably easier.

Question 4: We ask if there exists a positive constant s (maybe $s = 1$) with the property as follows. For each pair of simple closed planar curves γ and Γ and each homeomorphism $\mathcal{H}$ of γ onto Γ, there exists a homeomorphism $\hat{\mathcal{H}}$ of $\hat{\gamma}$ onto $\hat{\Gamma}$ with $\hat{\mathcal{H}}|\gamma = \mathcal{H}$ and

$$\max\{|\hat{\mathcal{H}}(x) - x| \, | \, x \in \hat{\gamma}\} \leq s \cdot \max\{|\mathcal{H}_h(x) - x| \, | \, x \in \gamma\}$$

where $\hat{\gamma} = \gamma \cap \text{int}\,(\gamma)$ and $\hat{\Gamma} = \Gamma \cap \text{int}\,(\Gamma)$.

Conjecture: [26]. Let Φ be Morse-Smale in Ω and inward on $\partial\Omega$. Then, for h sufficiently small, there exists a homeomorphism $\mathcal{J}_h : \Omega \to \mathcal{J}_h(\Omega) = \Omega \subset \mathbf{R}^n$ such that $|\mathcal{J}_h(x) - x| \leq \text{const} \cdot h^p$ for all $x \in \Omega$ and, preserving time-orientation, $\mathcal{J}_h$ maps each trajectory segment of Φ in Ω onto a $\varphi(h, \cdot)$-invariant curve in $\mathcal{J}_h(\Omega)$. The collection of these $\varphi(h, \cdot)$-invariant curves is identical to the system of trajectory curves of some local dynamical system Ψ_h on $\mathcal{J}_h(\Omega)$. In addition, if the rotation number of $\varphi(h, \cdot)$ on each Beyn-Eirola curve is irrational, then Ψ_h can be chosen to satisfy $\varphi(h, \cdot) = \Psi(h, \cdot)$.

The Conjecture is true in a tubular neighbourhood U of exponentially stable periodic orbits. The existence of positively $\varphi(h, \cdot)$-invariant curves near Γ_h follows from Theorem 1 when applied with $M = \Gamma$ and $M_h = \Gamma_h$. The rest is not hard. Diffeomorphism $\mathcal{D}_h$ gives automatically rise to a dynamics on the collection of these invariant curves in $\mathcal{D}_h(U \backslash \Gamma)$. In order to define Ψ_h on $\Gamma_h \cup \mathcal{D}_h(U \backslash \Gamma)$, one needs such a time-reparametrization on $\mathcal{D}_h(U \backslash \Gamma)$ that extends to a periodic motion on Γ_h. As it is explained in [24], any contractive toroidal coordinate system (θ, w) around Γ works for this purpose. In general, the resulting dynamics on Γ_h has nothing in common with $\varphi(h, \cdot)|\Gamma_h$. However, if the rotation number is irrational, then $\varphi(h, \cdot)|\Gamma_h$ embeds (as its time-h-map) into a continuous-time dynamical systems on Γ_h and the coordinate system induced by the numerical asymptotic phase applies.

2.4. The saddle structure about normally hyperbolic compact invariant manifolds is preserved under discretization. The first normal hyperbolicity result in numerical dynamics is due to Jens Lorenz [38]. He investigated stable tori. By using the standard 'one-chart' torus parametrization, he proved, in our notation, the existence of $\mathcal{F}_h$ and also inequality $\|\mathcal{F}_h - \text{inclusion}_\mathcal{M}\| \leq \text{const} \cdot h^p$. With Luca Dieci and Robert Russel [13], he developed numerical schemes for the computation of invariant tori, too. Though much has been done recently – see e.g. Gerard Moore [42] for later work – an effective and reliable computation of invariant tori is still not always possible.

Theorem 4: (case $k = 0$, $\mathcal{F}_h$-part) [22]; [28]. Let $\mathcal{M}$ be an eventually relative $(p + k + 1)$-normally hyperbolic (briefly: normally hyperbolic) compact invariant manifold with respect to $\dot{x} = f(x)$. Then there is a neighbourhood $\mathcal{N}$ of $\mathcal{M}$ in $\mathbf{R}^n$ such that, for h sufficiently small, the maximal compact invariant set for $\varphi(h, \cdot)$ in $\mathcal{N}$ is an invariant manifold $\mathcal{M}_h$, $\mathcal{M}_h = \mathcal{F}_h(\mathcal{M})$ for a C^{p+k+1} embedding $\mathcal{F}_h : \mathcal{M} \to \mathcal{N}$, $\mathcal{M}_h$ is normally hyperbolic with respect to $\varphi(h, \cdot)$ and the norm distance $\|\mathcal{F}_h - \text{inclusion}_{\mathcal{M}}\|_j$ in $C^j(\mathcal{M}, \mathbf{R}^n)$ is less than const $\cdot h^{\mu(j)}$, $j = 0, 1, \ldots, p + k$. In particular, $\mathcal{M}_h \to \mathcal{M}$ in the C^{p+k-1} norm topology on $C^{p+k+1}(\mathcal{M}, \mathbf{R}^n)$. Further, the global unstable manifold W_h^u of $\mathcal{M}_h$ is an injectively immersed copy of the global unstable manifold W^u of $\mathcal{M}$ and $W_h^u \to W^u$ in the C^{p+k-1} weak topology on $C^{p+k+1}(W^u, \mathbf{R}^n)$. The same result holds true for the corresponding global stable manifolds W_h^s and W^s.

The proof of Theorem 4 is a reconsideration of the Hirsch-Pugh-Shub approach [35] to normal hyperbolicity, with the necessary estimates supplemented. The argumentation goes along the chain $\Phi(h, \cdot) \longleftrightarrow \Phi(1, \cdot) \longleftrightarrow \varphi^N(h, \cdot) \longleftrightarrow \varphi(h, \cdot)$ where $N = N(h) \in \mathbf{N}$ is chosen in such a way that $Nh \in [1, 1 + h)$, $h \in (0, h_0]$. Existence and normal hyperbolicity of $\mathcal{M}_h$ – with respect to $\varphi^N(h, \cdot)$ first – follows directly from the Hirsch-Pugh-Shub results. (It is worth mentioning that, as a byproduct of these abstract results in the special case $\mathcal{M} = \Gamma$, the global unstable manifold W_h^u around Γ_h is $\varphi(h, \cdot)$-invariantly laminated by the injectively immersed C^{p+k+1} 'numerical asymptotic phase' submanifolds

$$W_{h,p}^{uu} = \left\{ x \in \mathbf{R}^n \,\|\varphi^{-l}(h, x) - \varphi^{-l}(h, p)| \to 0 \quad \text{as} \quad l \to \infty \right\}, \quad p \in \Gamma_h \,.)$$

The 'estimate'-part of the proof goes via comparing $\Phi(Nh, \cdot)$ and $\varphi^N(h, \cdot)$. The preliminary knowledge of the existence of the local unstable submanifold $W_{h,loc}^u$ is used in deriving the estimates and simplifies this task considerably.

Question 5: For h sufficiently small say $h \in [0, h_1]$, formula $\mathcal{F}(h, x) = \mathcal{F}_h(x)$ if $h \neq 0$ and x if $h = 0$ makes sense and defines a function $\mathcal{F} : [0, h_1] \times \mathcal{M} \to \mathbf{R}^n$. We ask if $\mathcal{F}$ is of class C^{p+k} (in the sense that it has a C^{p+k} extension $\hat{\mathcal{F}}$ defined on an open neighborhood of $[0, h_1] \times \mathcal{M}$ in $\mathbf{R} \times \mathcal{M}$).

We conjecture that the answer is affirmative and also that this affirmative answer is a consequence of the implicit function theorem. This is certainly the case when $\mathcal{M} = \{x_0\}$, a hyperbolic equilibrium point of $\dot{x} = f(x)$. Then $x = \mathcal{F}(h, x_0)$ solves the fixed point equation $\varphi(h, x) = x$ near x_0. Since $x = \varphi(0, x)$ and

$$\varphi(h, x) - x = \int_0^1 \varphi_h'(h\tau, x) d\tau \cdot h \quad \text{for all} \quad (h, x) \in [0, h_0] \times \mathbf{R}^n \,,$$

formula

$$\psi(h, x) = \begin{cases} h^{-1}(\varphi(h, x) - x) & \text{if } h \neq 0 \\ \varphi_h'(0, x) & \text{if } h = 0 \end{cases}$$

defines a C^{p+k} function $\psi : [0, h_0] \times \mathbf{R}^n \to \mathbf{R}^n$. For $h \neq 0$, our fixed point equation is equivalent to equation $\psi(h, x) = 0$. Observe that $\psi(h, x_0) = 0$ and $\psi'_x(0, x_0) = f'_x(x_0)$. Thus the implicit function theorem applies.

The amalgation of Theorems 3 and 4 leads to the following result. Let $p \geq 3$, $p + k \geq 4$ and assume that $\Phi|\mathcal{M}$ is Morse-Smale without periodic orbits. Then, for h sufficiently small, there exists a homeomorphism $\mathcal{K}_h$ of $\mathcal{M}$ onto $\mathcal{M}_h$ with the properties that $|\mathcal{K}_h(x) - x| \leq \text{const} \cdot h^{p-1}$ and $\mathcal{K}_h(\Phi(h, x)) = \varphi(h, \mathcal{K}_h(x))$ for all $x \in \mathcal{M}$.

2.5. Connecting orbits with transversal intersection are correctly reproduced by discretization. Mainly as a kind of boundary value problems, the doubly special case $j = 0$ and $\mathcal{M}, \tilde{\mathcal{M}}$ being hyperbolic equilibria and/or periodic orbits has been thoroughly investigated in the last years. For references, see [57].

Theorem 5: (case $\mathcal{M}, \tilde{\mathcal{M}}$ being equilibria) [25]; [28]. Let $\mathcal{M}$ and $\tilde{\mathcal{M}}$ be normally hyperbolic compact invariant manifolds with $\dim(W^u) + \dim(\tilde{W}^s) = n + 1$ and assume that $W^u \backslash \mathcal{M}$ and $\tilde{W}^s \backslash \tilde{\mathcal{M}}$ have a transversal intersection point P. Let $\gamma = \{\Phi(t, P) \in \mathbf{R}^n | t \in \mathbf{R}\}$. Then, for h sufficiently small, there exists an injectively immersed copy γ_h of γ with the properties that $\gamma_h \subset W_h^u \cap \tilde{W}_h^s$ and $\gamma_h \to \gamma$ in the C^{p+k-1} weak topology on $C^{p+k+1}(\gamma, \mathbf{R}^n)$.

Transversality at P means that the linear span of the tangent spaces of W^u and $\tilde{W}^s$ at P is the whole $\mathbf{R}^n$. It follows immediately that $\gamma = (W^u \backslash \mathcal{M}) \cap (\tilde{W}^s \backslash \tilde{\mathcal{M}})$ and that this intersection is, at each point of γ, transversal. This is a rank condition which makes the application of standard inverse function techniques along trajectory γ possible. Thus Theorem 5 is a direct consequence of Theorem 4. Observe that $\gamma_h = (W_h^u \backslash \mathcal{M}_h) \cap (\tilde{W}_h^s \backslash \tilde{\mathcal{M}}_h)$ for h sufficiently small and that this intersection is, at each point of the $\varphi(h, \cdot)$-invariant curve γ_h, transversal. If both $\mathcal{M}$ and $\tilde{\mathcal{M}}$ are hyperbolic periodic orbits, then the Haussdorff distance between γ and γ_h is bounded by $\text{const} \cdot h^{p-1}$.

The most interesting special case is when $\mathcal{M} = \tilde{\mathcal{M}} = \Gamma$, a hyperbolic periodic orbit. Then γ_h is homoclinic to the Beyn-Eirola curve Γ_h. In virtue of the famous Birkhoff-Smale theorem, the original differential equation $\dot{x} = f(x)$ is chaotic near Γ.

Question 6: We ask how this property is inherited by the discrete-time dynamical system $\{\varphi^M(h, \cdot)\}_{M \in \mathbf{Z}}$.

The difficulty is how to define Poincaré sections for $\varphi(h, \cdot)$ in such a way that the well-known C^1 perturbation theorems for topological horseshoes of diffeomorphisms could be applied. When combined with time-reparametrization, it seems plausible that $\varphi(h, \cdot)$ embeds in a continuous-time local dynamical system near $\Gamma \cup \gamma$ smoothly. This would lead to a proper counterpart of the classical Birkhoff-Smale theorem and would comply with the Conjecture, too.

A weaker (and also somewhat artificial) result is implied already by the last statement in the first paragraph of Subsection 2.1. This approach is outlined as follows.

Let S be a Poincaré section for the pair (Φ, Γ) and let $\mathcal{R} : S \supset Z \to S$ denote the corresponding first return map. For h sufficiently small, S is a transversal section of our continuous-time dynamical system Ψ_h whose solution curves fill a tubular neighborhood $\mathcal{U}$ of S. Projection $\mathcal{U} \to S$ along these solution curves is denoted by π_h. For $z \in Z$, choose $M = M(z) \in \mathbf{N}$ in such a way that Mh is near to $\tau(z) = \min\{\tau > 0 | \Phi(\tau, z) \in S\}$. By passing to a somewhat smaller Z if necessary, we may assume that $\varphi^M(h, z) \in \mathcal{U}$. Hence $\mathcal{R}_h(z) = \pi_h(\varphi^M(h, z))$ is defined and does not depend on the particular choice of $M = M(z)$. It is not hard to conclude that $\mathcal{R}_h$ is a small C^1-perturbation of $\mathcal{R}$. With a little care, S can be chosen in such a way that $\mathcal{R}$ satisfies the conditions of the Conley-Moser perturbation theorem; see e.g. in Wiggins [58]. The final conclusion is that, for h sufficiently small, certain iterates of the 'numerical first return map' $\mathcal{R}_h$ determine a topological horseshoe.

3 Facts behind

3.1. Function spaces and discretization on manifolds. Let M be an m-dimensional compact C^∞ manifold, $e : M \to \mathbf{R}^N$ a C^∞ (Whitney, for some $N \in \mathbf{N}$) embedding and let $\{(\alpha_i, U_i)\}_{i \in I}$ a finite coordinate atlas on M. By definition, $\{U_i\}_{i \in I}$ is an open cover of M and α_i is a C^∞ diffeomorphism mapping an open neighbourhood of cl (U_i) onto an open subset of $\mathbf{R}^m$. For $j \in \mathbf{N}$, let $C^j(M, M)$ and Diff$^j(M) \subset C^j(M, M)$ denote the set of C^j self-mappings and self-diffeomorphisms of M, respectively. Observe that Diff$^j(M)$ is an open subset of $C^j(M, M)$, $j \geq 1$. The linear space of C^j mappings of M to $\mathbf{R}^n$ is denoted by $C^j(M, \mathbf{R}^n)$. Given $F, G \in C^j(M, M)$ arbitrarily, formula

$$\max \{\sup \{|(eF\alpha_i^{-1} - eG\alpha_i^{-1})^{(r)}(s)||s \in \alpha_i(U_i)\}|r = 0, 1, \ldots, j; i \in I\} \tag{5}$$

defines $d_{C^j}(F, G)$ and thus a complete metric d_{C^j} on $C^j(M, M)$. Starting from another finite atlas and from another embedding, the resulting metrics are Lipschitz equivalent. Further, d_{C^0} is Lipschitz equivalent to the metric defined by $\max\{\rho(F(x), G(x))|x \in M\}$ on $C^0(M, M)$. Similarly, with embedding e omitted, formula (5) makes sense for $F, G \in C^j(M, \mathbf{R}^n)$ and defines a complete norm distance on $C^j(M, \mathbf{R}^n)$. Starting from another finite atlas, the resulting norms $\| \cdot \|_j$ are equivalent.

Definition: [26]. Let v be a C^{p+k+1} vector field on M. A C^{p+k+1} function $\varphi : [0, h_0] \times M \to M$ is a discretization mapping of order p if

$$d_{C^0}(\Phi(h, \cdot), \varphi(h, \cdot)) \leq \text{const} \cdot h^{p+1} \quad \text{for all} \quad h \in (0, h_0]$$

where $\Phi : \mathbf{R} \times M \to M$ is the solution flow of $\dot{x} = v(x)$, $x \in M$.

An important consequence of the Definition is the following counterpart of inequality (2). For h sufficiently small, $\varphi(h, \cdot)$ is a self-diffeomorphism of M and

$$d_{C^j}(\Phi(Mh, \cdot), \varphi^M(h, \cdot)) \leq \text{const} (|M|h) \cdot h^{\mu(j)},$$

for all $M \in \mathbf{Z}$ and $j = 0, 1, \ldots, p + k$.

Remark: With the notation we used in Theorem 4, define

$$\theta(h, x) = \begin{cases} \mathcal{F}_h^{-1}(\varphi(h, \mathcal{F}_h(x))) & \text{if } h \in (0, h_1], \ x \in \mathcal{M} \\ x & \text{if } h = 0, \ x \in \mathcal{M} \end{cases}$$

where $\mathcal{F}_h^{-1}$ denotes the inverse of $\mathcal{F}_h$ on $\mathcal{M}_h$, $h \in (0, h_1]$. It is easily seen that $d_{C^0}(\Phi(h, \cdot), \theta(h, \cdot)) \leq \text{const} \cdot h^p$ for all $h \in [0, h_1]$. Thus $\theta : [0, h_1] \times \mathcal{M} \to \mathcal{M}$ is a good candidate to be a discretization mapping (of order $p - 1$, with respect to the vector field $f|\mathcal{M}$). However, regardless of the suspected smoothness properties of θ (cf. Question 5 in Subsection 2.4), it is not hard to prove that

$$d_{C^j}(\Phi(h, \cdot), \theta(h, \cdot)) \leq \text{const} \cdot h^{\mu(j)}, \quad j = 0, 1, \ldots, p + k$$

and

$$d_{C^j}(\Phi(Mh, \cdot), \theta^M(h, \cdot)) \leq \text{const}(Mh) \cdot h^{\mu(j)-1}, \quad j = 0, 1, \ldots p + k - 1$$

for all $M \in \mathbf{N}$ and $h \in [0, h_1]$. This shows that the smoothness requirement in the Definition is not indispensable. (For example, inequality $d_{C^0}(\Phi(h, \cdot), \varphi(h, \cdot)) \leq \text{const} \cdot h^{p+1}$ is enough to imply that

$$d_{C^0}(\Phi(Mh, \cdot), \varphi^M(h, \cdot)) \leq \text{const}(Mh) \cdot h^p \quad \text{for all} \quad M \in \mathbf{N} \quad \text{and} \quad h \in [0, h_0]$$

no matter if φ is differentiable or not.) Nevertheless, as most discretization procedures are smooth, the smoothness requirement is incorporated in the Definition.

Next we explain the meaning of property "$W_h^u \to W^u$ in the C^{p+k-1} weak topology on $C^{p+k+1}(W^u, \mathbf{R}^n)$", one of the main assertions in Theorem 4. Assume that M is normally hyperbolic and consider its global unstable manifold W^u. In general, W^u is not a submanifold but – like the numeral 8 in $\mathbf{R}^2$ – only an immersed submanifold i.e. the image of an injective immersion of a submanifold. (A differentiable map is an immersion if, at each point of the domain, its derivative is injective.) For brevity, we say that S is a regular subset of W^u and $\{(\beta_i, V_i)\}_{i \in I}$ is a regular coordinate atlas on S if $M \subset S \subset W^u$, I is finite, $\{V_i\}_{i \in I}$ is an open cover of S in W^u, S and $N = \cup\{\text{cl}(V_i)|i \in I\}$ are compact, S is negatively Φ-invariant, M is the maximal invariant set in N, and β_i is a C^∞ diffeomorphism mapping an open neighbourhood of $\text{cl}(V_i)$ in W^u onto an open subset of some Euclidean space $\mathbf{R}^d$. Then there exists an injective C^{p+k+1} immersion $\mathcal{F}_h^u : W^u \to \mathbf{R}^n$ such that $W_h^u = \mathcal{F}_h^u(W^u)$ and, for each regular subset S of W^u with regular coordinate atlas $\{(\beta_i, V_i)\}_{i \in I}$

$$\max\{\sup\{|(\mathcal{F}_h^u \beta_i^{-1} - \text{inclusion}_{W^u} \beta_i^{-1})^{(r)}(s)|| s \in \beta_i(V_i)\}|i \in I\}$$

tends to zero (actually, is less than const $\cdot h^{\mu(r)}$) as $h \to 0^+$, $r = 0, 1, \ldots, p + k - 1$.

The definition of property "$\gamma_h \to \gamma$ in the C^{p+k-1} weak topology on $C^{p+k+1}(\gamma, \mathbf{R}^n)$" follows a similar pattern and is omitted.

3.2. Coordinate atlases with special properties. Theorem 4 concerns compact invariant manifolds with saddle-like behaviour nearby. The formal definition of normal hyperbolicity (see Definition 3 as well as Theorem 2.4 in [35]) is too elaborate to be given here in details. It requires the deployment of some fiber bundle concepts. However, the full richness of the definition lies within the existence of a particular coordinate atlas about M. This atlas consists of an infinite number of exponential bundle charts in which a parametrized version of the contraction mapping principle in differential topology (the so-called graph transformation method) works. As for an analogy, recall the first step in proving Grobman-Hartman Lemma. This is the introduction of a new norm expressing contracting-expanding behaviour in such a way that is eminently suitable for computations – in our present terminology, the introduction of a single coordinate chart with special properties. Also the proof of Theorem 3 is using exponential charts.

The rest of this section is devoted to some interesting properties of composition and inversion operators which also play a not unimportant role in proving Theorems 3 and 4.

3.3. Some differential-topological properties of composition and inversion. Fix $g \in C^{r+s}(M, M)$ with some $r, s \in \mathbf{N}$. For $q = 0, 1, \ldots, r$, the left composition operator $f \to g \circ f$ maps $C^r(M, M)$ in $C^{r-q}(M, M)$ and is of class C^{s+q}. The right composition operator $f \to f \circ g$ is less interesting: it is a continuous linear self-mapping of $C^q(M, M)$, $q = 0, 1, \ldots, r + s$. With both $g \in C^{r+s}(M, M)$ and $f \in C^{r-q}(M, M)$ varying, operator $(g, f) \to g \circ f$ maps $C^{r+s}(M, M) \times C^r(M, M)$ in $C^{r-q}(M, M)$ and is of class C^{s+q}, $q = 0, 1, \ldots, r$. Similarly, the inversion operator $f \to f^{-1}$ maps $\text{Diff}^r(M)$ in $\text{Diff}^{r-q}(M)$ and is of class C^q, $q = 0, 1, \ldots, r$.

It is Lipschitz consequences of these facts plus inequality (2) that lie beyond the frequent appearance of the "magic" exponents $\mu(0) = p$, $\mu(j)$, $\mu(j) + 1$ and $\mu(j + 1)$ in all inequalities of Section 2 above.

While working on coordinate chart representations, inequalities

$$|(ab - AB)^{(j)}| \leq K \cdot \max\{|a^{(k)} - A^{(k)}| + |b^{(k)} - B^{(k)}| | k = 0, 1, \ldots, j\}$$

and

$$|(c^{-1} - C^{-1})^{(j)}| \leq L \cdot \max\{|c^{(k)} - C^{(k)}| | k = 0, 1, \ldots, j\}$$

are particularly useful. Here $a, A, c, C \in BC^{j+1}$ and $b, B \in BC^j$ are functions an open subset of $\mathbf{R}^d$ we do not specify (letter B preceding C = "continuity" stands for "boundedness") and K resp. L is a polynomial function of the norms $\{|a^{(k)}|, |b^{(k)}|, |B^{(k)}|\}_{k=1}^j$

and $\{|A^{(k)}|\}_{k=1}^{j+1}$ resp. $\{|c^{(k)}|, |(c^{-1})^{(k)}|\}_{k=1}^{j}$ and $\{|C^{(k)}|, |(C^{-1})^{(k)}|\}_{k=1}^{j+1}$. It is of course assumed that the inverse functions exist and $c^{-1}, C^{-1} \in BC^{j+1}$. This can be ensured by requireing that the linear operator C^{-1} is invertible and $|c - C|$ and $|c^{-1} - C^{-1}|$ are sufficiently small. Estimates for the size of the respective domains as well as for the norms $\{|(c^{-1})^{(k)}|, |(C^{-1})^{(k)}|\}_{k=1}^{j+1}$ are also at hand.

3.4. *Some combinatorical properties of composition and inversion.* The proof of inequality (2) is based on simple Pascal triangle properties of chain rule formulae for $(\varphi^M(h, \cdot))_x^{(j)}$. More precisely, a rather lengthy but straightforward inductive-enumerative application of the classical/Pascal binomial theorem shows that (2) is a consequence of (1) and (4). Starting from the elementary inequality

$$|(\Phi(h, x) - \varphi(h, x))_{h,x}^{(i,j)}| \leq \text{const} \cdot h^{\mu(i+j)-i+1}$$

similar considerations for mixed partial derivatives yield also that

$$|(\Phi(Mh, x) - \varphi^M(h, x))_{h,x}^{(i,j)}| \leq \text{const} (|M|h) \cdot h^{\mu(i+j)-i} \tag{6}$$

for all $(h, x) \in [0, h_0] \times \mathbf{R}^n$, $M \in \mathbf{Z}$ and $i, j \in \mathbf{N}$, $\mu(i + j) \geq i$, $i + j = 0, 1, \ldots, p + k$. In case of $\mathcal{L} \geq 0$, it is not hard to show that, given $\varepsilon > 0$ arbitrarily,

$$\text{const}(Mh) \leq c_\varepsilon \cdot \exp\left((\mathcal{L} + \varepsilon)(i + j + 1)Mh\right), \quad M \in \mathbf{N}.$$

For $j = 0$ (and even for $\mu(i) \leq i$), as it is shown by the one-dimensional linear example $\Phi(h, x) = \exp(h)x$ with Euler method $\varphi(h, x) = x + hx$, exponent $\mu(i) - i$ in (6) is sharp. In fact, with $hM \to 1$ and $h \to 0$, a direct computation shows that

$$\lim\{h^{i-1}(e^{Mh} - (1 + h)^M)_h^{(i)}\} = 2^{-1}(1 + i + i^2)e, \quad i \in \mathbf{N}.$$

Also Abel's version of the binomial theorem can be used. For example, with F denoting a C^∞ real function and F^a the a-th power (not self-composition!) of F, it holds true that

$$\sum_{k=0}^{n} \binom{n}{k}(k + \lambda)^{-1}(F^{k+\lambda})^{(k)}(n + \mu - k)^{-1}(F^{n+\mu-k})^{(n-k)}$$

$$= (\lambda^{-1} + \mu^{-1})(n + \lambda + \mu)^{-1}(F^{n+\lambda+\mu})^{(n)} \quad \text{whenever} \quad n, \lambda, \mu \in \mathbf{N}^+.$$

(With $F = \exp$, the latter formula simplifies to one of Abel's identities.) The case $\lambda = \mu = 1$ is of particular interest. It results in the formal expansion

$$(\text{id} - F)^{-1} = \text{id} + \sum_{k=1}^{\infty}(k!)^{-1}(F^k)^{(k-1)}$$

and implies also that

$$u(t, x) = F((\text{id} - tF)^{-1}(x)) = \sum_{k=0}^{\infty}((k + 1)!)^{-1}t^k(F^{k+1})^{(k)}(x), \quad (t, x) \in [0, \infty) \times \mathbf{R}$$

is a formal solution of Burger's initial value problem $u_t' = uu_x'$, $u(0, \cdot) = F$.

4 The real problems, on their way of being solved

It is much harder to decide if the solution flow of a particular equation has a certain qualitative property than to construct a dynamical system with a prescribed geometric behaviour. Instead of the qualitative properties of $\dot{x} = f(x)$, we are usually given some qualitative properties of a numerical approximation/simulation. Does it follow that the original ODE possesses the same qualitative properties? More generally, assume that numerical/computer simulations have led to certain conjectures on the qualitative behaviour of a particular continuous-time or discrete-time dynamical system. Can such conjectures be proved rigorously? These are reverse perturbation problems where also the effect of rounding errors has to be taken into consideration.

Assume that inequality (0) is replaced by

$$|\Phi(h, x) - \tilde{\varphi}(h, x)| \leq ch^{p+1} + \tau \quad \text{for all} \quad h \in [0, h_0], \quad x \in \mathbf{R}^n$$

where $\tilde{\varphi}(h, x) = \varphi(h, x) + \text{error}(h, x)$ and c, τ are nonnegative constants, τ small. Depending on the finer structure of equation $\dot{x} = f(x)$, the standard error estimate

$$|\Phi(Mh, x) - \tilde{\varphi}^M(h, x)| \leq (\mathcal{L} + \varepsilon)^{-1} \cdot \exp((\mathcal{L} + \varepsilon)Mh) \cdot (ch^p + \tau/h)$$

(valid for all $(h, x) \in [0, h_0] \times \mathbf{R}^n$ and $M \in \mathbf{N}$ where $\mathcal{L} + \varepsilon$ is taken from inequality (4) and assumed to be positive) can often be considerably improved. Under the conditions of Theorem 3, there is a positive constant C (independent of h and τ) and, for all h sufficiently small, there exists a not necessarily continuous or invertible mapping $\mathcal{K}_h : \Omega \to \mathbf{R}^n$ such that

$$|\Phi(Mh, \mathcal{K}_h(x)) - \tilde{\varphi}^M(h, x)| \leq C(ch^p + \tau/h) \quad \text{whenever} \quad x \in \Omega \quad \text{and} \quad M \in \mathbf{N}.$$

According to their Remark (ii), this is a special case of Theorem 2.3 of Shui-Nee Chow and Erik Van Vleck [12]. The geometric meaning is that numerical solutions are shadowed (uniformly, on the whole time-set $\{Mh\}_{M\in\mathbf{N}}$) by exact solutions. The $\tau = 0$ case is already settled by Theorem 3 which implies that

$$|\Phi(Mh, x) - \varphi^M(h, \mathcal{J}_h(x))| = |\Phi(Mh, x) - \mathcal{J}_h(\Phi(Mh, x))| \leq \text{const} \cdot h^p$$

for all $x \in \Omega$ and $M \in \mathbf{N}$, a result (because $\mathcal{J}_h$ is a homeomorphism) on mutual shadowing. This is not bad at all but we can hardly know a priori that the conditions of Theorem 3 are satisfied.

What we are usually given is only a finite sequence of numerical iterates $\{\tilde{\varphi}^M(h_0, x_0)\}_{M=0}^N$. However, if one is lucky enough in choosing the starting point x_0 – and one is often lucky especially near to hyperbolic sets suspected –, a numerical investigation of the variational equation by computer can point out a sort of exponential dichotomy along $\{\tilde{\varphi}^M(h_0, x_0)\}_{M=0}^N$ and can lead to a shadowing assignment $\kappa(x_0)$ with the rigorous error estimate

$$|\Phi(Mh_0, \kappa(x_0)) - \tilde{\varphi}^M(h_0, x_0)| \leq C(f, x_0)(ch_0^p + \tau/h_0) \quad \text{whenever} \quad M = 0, 1, \ldots, N.$$

The above considerations summarize biefly several papers by James Yorke as well as Kenneth Palmer and their associates on long-term path-following algorithms. The point is that constant $C(f, x_0)$ might be much-much smaller than $(\mathcal{L} + \varepsilon)^{-1}\exp((\mathcal{L} + \varepsilon)Nh_0)$ and that $C(f, x_0)$ is 'constructed' by the computer during the numerical procedure. Moreover, still within the broad framework of exponential dichotomy, the method can be used for computer-aided establishing the existence of periodic trajectories. Characteristic representatives of this development are e.g. [47], [53].

The most exciting results of computer-assisted dynamical systems theory relate to rigorous chaos verification. The high-point is the work of Konstantin Mischaikow and Marian Mrozek [40], [41] who created a general method for proving the existence of horseshoe-type chaos by computer. The underlying abstract results are perturbation theorems of algebraic topology establishing Conley-Moser-like geometric criteria for shift dynamics. These geometric criteria – the presence of the 'skeleton' of a geometric horseshoe on a transversal section – have already been checked by computer for Lorenz and Rössler equation, with a wide range of parameter values including the classical parameters. Further, as it was pointed out by Piotr Zgliczynsky [59], the seventh iterate of Henon's map (with the classical parameter values) has a topological horseshoe. Beautiful, isn't it?

For details, see the contributions of Marian Mrozek [44] and Roman Srzednicki [55] in this volume surveying recent results and demonstrating the role of the Wazewski school – named after the distinguished Polish mathematician Tadeusz Wazewski(1896-1972), late professor of the Jagiellonian University in Cracow – in the development of the subject.

The original approach of [40] has been considerably simplified in the meantime. It has turned out that discrete Conley index [43] can be replaced by traditional categories of algebraic topology including fixed point index. Also computing time has been considerably reduced by using numerical algorithms that fit better to interval arithmetics (see the references in Oliver Abeth [1]), the theory of rigorous estimation of computer errors. It is a must to refer here to an alternative approach as well. Its essence – a rigorous computer search for suspected intersection points between stable and unstable manifolds – manifests itself in the title of a paper by Brian Hassard and Jianhe Zhang [33]: 'Existence of a homoclinic orbit of the Lorenz system by precise shooting'. We can well imagine that, in a few years or so, chaos-checking computer programs become part of standard software. For an introductory paper, we recommend Arnold Neumaier and Thomas Rage [46]. For details and further development, see an avelanche of papers to come.

Very recently, Xinfu Chen [11] has given the first analytical proof for the existence of a transverse homoclinic point in the Lorenz equation (still not for the classical parameter values). To the best of our knowledge, the existence of a strange attractor is still unknown. On the other hand, the existence of a strange attractor for the Henon map is already known (for a set of parameter values of positive Lebesgue measure, see e.g. in Mark Pollicott [49]).

5   Appendix

Concluding this paper, we return to the differential-topological properties of composition and inversion discussed in the first paragraph of Subsection 3.3. Results on the composition operator go back to Smale, see Abraham [2]. Their proofs (with slightly different order of generality) can be found also in Abraham, Marsden and Ratiu [3], Foster [19], Franks [20] and Irwin [36]. The one-dimensional case is throughoutfully discussed in the monograph of Appell and Zabreiko [4].

With $C^\infty(M, M)$ carrying an appropriate Fréchet structure, operator $(g, f) \to g \circ f$ maps $C^\infty(M, M) \times C^\infty(M, M)$ in $C^\infty(M, M)$ and is of class C^∞. Similarly, operator $f \to f^{-1}$ maps $C^\infty(M, M)$ in $C^\infty(M, M)$ and is of class C^∞. These two latter facts belong to the fundamentals of infinite-dimensional Lie group theory, see e.g. Section 2 in Ebin and Marsden [15] or Milnor [39]. We are indebted to Prof. J. Szenthe for his kind help in setting up the above list of references.

However, contrary to the case of composition operators, differentiability results on the operator $f \to f^{-1}$ between spaces of finite degree of smoothness do not seem to be widely known. It is usually remarked in surveys on map spaces that $f \to f^{-1}$ as a self-map of $\text{Diff}^1(M)$ is not Fréchet differentiable. We erroneously stated in a half-sentence remark accompanying (but wholly independent of) the proof of Theorem 3.8 in [26] that it was of class C^∞. Actually, as we shall show below, it is nowhere Fréchet differentiable. This is the first aim of the present Appendix. The second aim is to investigate the inversion operator as a mapping from $\text{Diff}^r(M)$ to $\text{Diff}^{r-q}(M)$, $q = 0, 1, 2, \ldots, r$. Our lemma below is surely known. Its proof goes along the standard route of obtaining differentiability results for the composition operator. Nevertheless, we could not find it in the literature. Its presentation as a separate Lemma is thus not without any reason.

Proposition a.) Operator $\varphi \to \varphi^{-1}$ is a self-homeomorphism of $\text{Diff}^1(M)$ but b.) it is not Fréchet differentiable at any point.

Proof. a.) This is a well-known consequence of the inverse function theorem.

b.) Suppose not. Working on coordinate chart representations, there exists a C^1 diffeomorphism f and a bounded linear map $L : BC^1 \to BC^1$ such that

$$\|(f + h)^{-1} - f^{-1} - Lh\|_1 = o(\|h\|_1) \quad \text{for} \quad h \in BC^1 \quad \text{small} \tag{7}$$

where BC^1 is the vector space of $\mathbf{R}^m$-valued bounded C^1 functions (defined on an unspecified bounded open subset of $\mathbf{R}^m$) with bounded first derivative and $\|h\|_0 = \sup\{|h(x)| \,|\, x \in \text{domain}\}$, $\|h\|_1 = \max\{\|h\|_0, \|h'\|_0\}$. There is no loss of generality in assuming that $(f^{-1})' = [f'(f^{-1})]^{-1}$ is uniformly continuous.

Next we show that operator $\varphi \to \varphi^{-1}$ as a mapping of $\text{Diff}^1(M)$ in $C^0(M, M)$ is differentiable and (in a coordinate chart representation) that its derivative at f is the bounded linear map $BC^1 \to BC^0$, $h \to -[(f^{-1})']h(f^{-1})$. Thus we have to check that

$$\|(f + h)^{-1} - f^{-1} + [(f^{-1})']h(f^{-1})\|_0 = o(\|h\|_1) \quad \text{for} \quad h \in BC^1 \quad \text{small.} \tag{8}$$

In fact, with $y = (f + h)^{-1}(x)$, an elementary computation yields that

$$|f^{-1}(f(y)) - f^{-1}(f(y) + h(y)) + [(f^{-1})'(f(y) + h(y))]h(f^{-1}(f(y) + h(y)))|$$

$$= \left| \int_0^1 [(f^{-1})'(f(y) + h(y)) - (f^{-1})'(f(y) + \tau h(y))]h(y)d\tau + \right.$$

$$\left. [(f^{-1})'((f(y) + h(y))]\{h(y) - h(f^{-1}(f(y) + h(y)))\} \right|$$

$$\leq o(\|h\|_0) + \text{const} \cdot \|h'\|_0 \cdot |f^{-1}(f(y)) - f^{-1}(f(y) + h(y))|$$

$$= o(\|h\|_0) + \text{const} \cdot \|h'\|_0 \cdot \left| \int_0^1 [(f^{-1})'(f(y) + \tau h(y))]h(y)d\tau \right| = o(\|h\|_1)$$

for $h \in BC^1$ small. Taking supremum in y, (8) follows.

Comparing (7) and (8), uniqueness of the derivative yields that

$$\text{incl}_{BC^1 \to BC^0}(Lh) = -[(f^{-1})']h(f^{-1}) .$$

In particular, $[(f^{-1})']h(f^{-1}) \in BC^1$ for $h \in BC^1$, $\|h\|_1$ small. Consider now the matrix representation $\{q_{i,j}(x)\}_{i,j=1}^m$ of $[(f^{-1})'(x)]$. Varying r with $h = r(f)$ in BC^1, we see via Cramer's rule that $q_{ij} \in BC^1$, $i, j = 1, \dots, m$. By passing to the inverse matrix, we end up with the C^1 property for $[(f^{-1})']^{-1} = f'(f^{-1})$. The final conclusion is – via $f' = (f'(f^{-1}))f$ – that f is of class C^2. By a slight restriction of the domain, we may assume that f'' is uniformly continuous and $|f'|, |f''| \leq K$ for some constant K.

For brevity, set $B = (f' + h')(f + h)^{-1}$ and $A = f'(f^{-1})$. We claim that

$$\max\{\|B^{-1} - A^{-1}\|_0, \|B - A\|_0\} \leq \text{const} \cdot \|h\|_1 \quad \text{for} \quad h \in BC^1 \quad \text{small.} \tag{9}$$

As a simple consequence of (8), $\|(f+h)^{-1} - f^{-1}\|_0 \leq \text{const} \cdot \|h\|_1$. It follows easily that $\|B - A\|_0 \leq K\|(f+h)^{-1} - f^{-1}\|_0 + \|h'(f+h)^{-1}\|_0 \leq \text{const} \cdot \|h\|_1$. On the other hand, the resolvent identity $B^{-1} - A^{-1} = -A^{-1}(B-A)B^{-1}$ implies that $\|B^{-1} - A^{-1}\|_0 \leq \text{const} \cdot \|B - A\|_0$ and (9) follows.

Now we are in a position to simplify (7). Differentiating $(f+h)^{-1} - f^{-1} - Lh$, we see that (7) is equivalent to

$$\|B^{-1} - A^{-1} + [(A^{-1})']h(f^{-1}) + A^{-1}h'(f^{-1})A^{-1}\|_0 = o(\|h\|_1) \quad \text{for} \quad h \in BC^1 \quad \text{small.}$$

Since $B^{-1} - A^{-1} = -A^{-1}(B-A)A^{-1} + (A^{-1} - B^{-1})(B-A)A^{-1}$ and

$$[(A^{-1})']h(f^{-1}) = -A^{-1}f''(f^{-1})A^{-1}(A^{-1}h(f^{-1})) = -A^{-1}f''(f^{-1})(A^{-1}h(f^{-1}))A^{-1} ,$$

inequality (9) implies that (7) is equivalent to

$$\| - (f' + h')(f+h)^{-1} + f'(f^{-1}) - f''(f^{-1})A^{-1}h(f^{-1}) + h'(f^{-1})\|_0 = o(\|h\|_1)$$

for $h \in BC^1$ small. A twofold application of (8) plus the uniform continuity of f'' yield that

$$\|f'(f+h)^{-1} - f'(f^{-1}) + f''(f^{-1})A^{-1}h(f^{-1})\|_0 =$$

$$\| \int_0^1 \{f''(f^{-1} + \tau((f+h)^{-1} - f^{-1})) - f''(f^{-1})\} \cdot ((f+h)^{-1} - f^{-1})d\tau$$

$$+ f''(f^{-1}) \cdot \{(f+h)^{-1} - f^{-1} + A^{-1}h(f^{-1})\}\|_0 = o(\|h\|_1)$$

and consequently, (7) is equivalent to

$$\|h'(f+h)^{-1} - h'(f^{-1})\|_0 = o(\|h\|_1) \quad \text{for} \quad h \in BC^1 \quad \text{small.} \tag{10}$$

With h replaced by λs, for some $\lambda \in \mathbf{R}$ and s in $\mathcal{S} = \{s \in BC^1 \mid \|s\|_1 = 1\}$, (10) can be rewritten as

$$\sup\{\|s'(f+\lambda s)^{-1} - s'(f^{-1})\|_0 \mid s \in \mathcal{S}\} \to 0 \quad \text{whenever} \quad \lambda \to 0 . \tag{11}$$

Given $T > 0$ arbitrarily, define $\mathcal{B}_T = \{s \in BC^1 \mid \|s\|_1 \leq T\}$. As a trivial consequence of (11), it holds also true that

$$\sup\{\|s'(f+\lambda s)^{-1} - s'(f^{-1})\|_0 \mid s \in \mathcal{B}_T\} \to 0 \quad \text{whenever} \quad \lambda \to 0 \tag{12}$$

for each $T > 0$ separately.

We claim that (12) is impossible, even at $x = x_0$ arbitrarily fixed. In fact, there is no loss of generality in assuming that f is defined for all $v \in \mathbf{R}^m$ with $|v| < 1$ and $f^{-1}(x_0) = 0$. In virtue of (8),

$$\lambda^{-1} \cdot \sup\{\|(f+\lambda s)^{-1} - f^{-1} + \lambda[(f^{-1})']s(f^{-1})\|_0 \mid s \in \mathcal{B}_T\} \to 0 \quad \text{whenever} \quad \lambda \to 0 ,$$

for each $T > 0$ separately. In particular,

$$\lambda^{-1} \cdot \sup\{|(f + \lambda s)^{-1}(x_0) + \lambda[(f'(0)]^{-1}s(0)||s \in \mathcal{B}_T\} \to 0 \quad \text{whenever} \quad \lambda \to 0 \,,$$

for each $T > 0$ separately. Let $\alpha > \max\{1, |[f'(0)]^{-1}|^{-1}\}$ and pick a $d \in \mathbf{R}^m$ with $|d| = (2\alpha)^{-1}$. For each $N = 1, 2, \ldots$, it is not hard to construct a C^∞ function $s_N \in \mathcal{B}_2$ with the properties that $[f'(0)]^{-1}s_N(0) = d$, $|s_N(0)| < 1/2$, $s'_N(0) = 0$ and, last but not least,

$$|s'_N(v)| > 1 \quad \text{whenever} \quad |v + N^{-1}d| < (4\alpha N)^{-1} \,.$$

It follows that

$$N|(f + N^{-1}s_N)^{-1}(x_0) + N^{-1}d| < (4\alpha)^{-1}$$

and consequently,

$$|s'_N((f + N^{-1}s_N)^{-1}(x_0)) - s'_N(0)| > 1 \quad \text{for } N \text{ sufficiently large.}$$

But this contradicts (12) and thus concludes the proof of Proposition b.). However, some little more care is needed. We have namely to ensure that our functions $f + h$ (i.e. $f + r(f)$ and $f + \lambda s$ (in particular $f + N^{-1}s_N$)) are coordinate chart representations of self-diffeomorphisms of M. This latter requirement relates to (7), (10), (11) and (12), too. The coordinate chart constructions can be lifted to M and our functions, constituting local diffeomorphisms on M, can be extended to self-diffeomorphisms of M.

   Example: (In collaboration with Prof. G. Petruska.) The operator we discussed in Proposition b.) is not the simplest example for nowhere differentiable homeomorphisms in infinite dimension. In fact, with ℓ_1 denoting the usual Banach space of real sequences $x = (x_1, x_2, x_3, \ldots)$ with norm $\|x\| = \sum |x_n|$, consider the mapping

$$\mathcal{H} : \ell_1 \to \ell_1, \quad x \to (\|x\| + 2x_1, x_2, x_3, \ldots) \,.$$

It is elementary to check that $\mathcal{H}$ is a self-homeomorphism of ℓ_1 and $\mathcal{H}^{-1}(x) = (\alpha(x), x_2, x_3, \ldots)$ where $\alpha = \alpha(x) \in \mathbf{R}$ is the unique solution of equation

$$|\alpha| + 2\alpha = x_1 - |x_2| - |x_3| - \ldots, \quad \text{for each} \quad x \in \ell_1 \,.$$

But the norm function $\| \cdot \| : \ell_1 \to \mathbf{R}^+$, $x \to \|x\|$ is nowhere differentiable. The easy proof can be found e.g. in Phelps [48,p.8]. The final conclusion is that $\mathcal{H}$ is not differentiable at any point of ℓ_1. Alternatively, let $j : \mathbf{R} \to \mathbf{R}$ be a nowhere differentiable continuous function. For $(x_1, x_2) \in \mathbf{R}^2$, define $\mathcal{J}(x_1, x_2) = (j(x_1) + x_2, x_1)$. It is obvious that $\mathcal{J}$ is nowhere differentiable. Further, $\mathcal{J}$ is a self-homeomorphism of $\mathbf{R}^2$. This two-dimensional example gives rise to a similar example in all Banach spaces (of dimension ≥ 2) including ℓ_1. The speciality of $\mathcal{H}$ is its Lipschitz property. We ask if such a Lipschitz example exists in other (which?) infinite-dimensional Banach spaces – $\mathbf{R}^n$ is excluded by the classical Rademacher theorem.

Remark: Returning to the proof of Proposition b.), let S be an arbitrary subset of $\mathcal{S}$. Assume that f is of class C^2. The equivalence of (7) and (11) shows that the convergence properties

$$\lambda^{-1} \cdot \sup \{\|(f + \lambda s)^{-1} - f^{-1} + \lambda[(f^{-1})']s(f^{-1})\|_1 | s \in S\} \to 0 \ \text{ as } \ \lambda \to 0 \qquad (13)$$

(observe that $S = \mathcal{S}$ resp. $S = \{\text{singletons}\}$ in (13) means standard Fréchet resp. Gateaux differentiability) and

$$\sup \{\|s'(f + \lambda s)^{-1} - s'(f^{-1})\|_0 | s \in S\} \to 0 \ \text{ as } \ \lambda \to 0 \qquad (14)$$

are also equivalent. By a standard Arzela-Ascoli argument, (14) holds true provided that S is compact in $\mathcal{S} \subset BC^1$. Using a terminology which goes back to Hadamard, this means that operator $\varphi \to \varphi^{-1}$ as a self-homeomorphism of $\text{Diff}^1(M)$ is compactly differentiable at all points of $\text{Diff}^2(M) \subset \text{Diff}^1(M)$. Since BC^2 is compactly embedded in BC^1, it follows also that $\varphi \to \varphi^{-1}$ as a mapping of $\text{Diff}^2(M)$ in $\text{Diff}^1(M)$ is (Fréchet) differentiable. In spaces of distribution functions on the unit interval or the real line, compact differentiability of the inversion operator at certain special distributions has been known for some time and plays a role in probability theory. For details, see Dudley [14]. We are indebted to Prof. J. Appell for pointing out this reference.

Lemma: The inversion operator $\varphi \to \varphi^{-1}$ maps $\text{Diff}^r(M)$ in $\text{Diff}^{r-q}(M)$ and is of class C^q, $q = 0, 1, \ldots, r$.

Proof. Case $r = 0$ follows from an easy compactness argument. From now on assume that $r \geq 1$. Case $q = 0$ is a well-known consequence of the inverse function theorem.

The next step is to settle case $q = 1$. As in proving Proposition b.), we pass to coordinate chart representations. What we have to prove is that

$$\|(f + h)^{-1} - f^{-1} + [(f^{-1})']h(f^{-1})\|_{r-1} = o(\|h\|_r) \ \text{ for } \ h \in BC^r \ \text{ small} \qquad (15)$$

where f is a diffeomorphism in BC^r, the vector space of $\mathbf{R}^m$-valued C^r functions with all derivatives bounded and equipped with norm $\|h\|_r = \max\{\|h\|_0, \|h'\|_0, \ldots, \|h^{(r)}\|_0\}$. Property (15) is derived from (8) via induction on r and can be found in Hamilton [32, p.92].

Like higher order versions of the classical C^1 inverse function theorem, the $q \geq 2$ case of the Lemma follows now from the formula for the first derivative. For simplicity, we write $Qf = f^{-1}$. Our aim is to show that $Q \in C^q(BC^r, BC^{r-q})$ or equivalently, that $DQ \in C^{q-1}(BC^r, L(BC^r, BC^{r-q}))$. This can be done via induction on q. (Here, of course, $L(BC^r, BC^{r-q})$ stands for the Banach space of bounded linear operators from BC^r in BC^{r-q}.) Recall that

$$\{DQ(f)\}(h) = -[f' \circ f^{-1}]^{-1} \cdot (h \circ f^{-1}) \ \text{ for all } \ h \in BC^r \ .$$

As a composition of C^{q-1} mappings (including the two-variable composition operator itself; see the first paragraph of Subsection 3.3), operator

$$BC^r \to BC^{r-q}(\cdot, L(\mathbf{R}^m, \mathbf{R}^m)), \quad f \to \eta = -[f' \circ f^{-1}]^{-1}$$

is of class C^{q-1}. It remains thus to prove that operator

$$BC^r \times BC^{r-q}(\cdot, L(\mathbf{R}^m, \mathbf{R}^m)) \to L(BC^r, BC^{r-q}), \quad (f, \xi) \to (h \to \xi \cdot (h \circ f^{-1}))$$

or equivalently (due to linearity in ξ), operator

$$BC^r \to L(BC^r, BC^{r-q}), \quad f \to (h \to h \circ f^{-1})$$

is also of class C^{q-1}. Since operator $BC^r \to BC^{r-q}$, $f \to f^{-1}$ is assumed to be of class C^{q-1}, we have to show lastly that operator

$$BC^{r-q} \to L(BC^r, BC^{r-q}), \quad g \to (h \to h \circ g)$$

is also of class C^{q-1}. But this is exactly Lemma B.14 of Irwin [36] and so the proof is completed.

Question: We ask if $\varphi \to \varphi^{-1}$ as a self-homeomorphism of $\mathrm{Diff}^r(M)$, $r \geq 2$ is nowhere Fréchet differentiable. Similarly, we ask if it is compactly differentiable at all points of $\mathrm{Diff}^{r+1}(M) \subset \mathrm{Diff}^r(M)$, $r \geq 2$. Both answers seem to be affirmative.

References

1. Abeth, O. (1994): Computation of topological degree using interval arithmetics, and applications. Math. Comput. **62**, 171-178

2. Abraham, R. (1963): Lectures of Smale on Differential Topology. Columbia University (mimeographed notes, unpublished)

3. Abraham, R., Marsden, J.E., Ratiu, T. (1983): Tensor Analysis and Applications. Springer, Berlin

4. Appell, J., Zabreiko, P.P. (1990): Nonlinear Superposition Operators. Cambridge University Press, Cambridge

5. Aulbach, B., Garay, B.M. (1994): Discretization of semilinear differential equations with an exponential dichotomy. Computers Math. Appl. **28**, 23–35

6. Beyn, W.J. (1987a): On invariant curves for one-step methods. Numer. Math. **51**, 103–122

7. Beyn, W.J. (1987b): On the numerical approximation of phase portraits near stationary points. SIAM J. Numer. Anal. **24**, 1095–1113

8. Beyn, W.J., Lorenz, J. (1987): Center manifolds of dynamical systems under discretization. Numer. Funct. Anal. Optimiz. **9**, 381–414.

9. Butcher, J.C. (1987): The Numerical Analysis of Ordinary Differential Equations. Wiley, Chichester

10. Calvo, M.P., Sanz-Serna, J.M. (1994): Numerical Hamiltonian Problems. Applied Mathematics and Mathematical Computation **7**, Chapman and Hall

11. Chen, X.Y. (1995): Lorenz equations, Part III: existence of hyperbolic sets. (preprint)

12. Chow, S.N., Van Vleck, E. (1994): A shadowing lemma approach to global error analysis for initial value ODE's. SIAM J. Sci. Comput. **15**, 959–976

13. Dieci, L., Lorenz, J., Russel, R.D. (1991): Numerical calculation of invariant tori. SIAM J. Sci. Stat. Comput. **12**, 607–647.

14. Dudley, R.M. (1994): The order of the remainder in derivation of composition and inverse operators for p-variation norms. Ann. Stat. **22**, 1–20

15. Ebin, D.G., Marsden, J.E. (1970): Groups of diffeomorphs and the motion of an incompressible fluid. Ann. Math. **92**, 102–163.

16. Eirola, T. (1988): Invariant curves of one-step methods. BIT **28**, 113–122

17. Eirola, T. (1993): Aspects of backward error analysis in numerical ODE's. J. Comput. Appl. Math. **45**, 65–73

18. Fečkan, M. (1992): The relation between a flow and its discretization. Math. Slovaca **42**, 123–127

19. Foster, M.J. (1975): Calculus on vector bundles. J. London. Math. Soc. **11**, 65–73

20. Franks, J. (1979): Manifolds of C^r mappings and applications to differentiable dynamical systems. In: G.C. Rota ed., Studies in Analysis, Advances in Mathematics Supplementary Studies **4**, pp. 271–290. Academic Press, New York

21. Garay, B.M. (1993): Discretization and some qualitative properties of ordinary differential equations about equilibria. Acta Math. Univ. Comenianae **62**, 249–275

22. Garay, B.M. (1994a) Discretization and normal hyperbolicity. Z. angew. Math. Mech. **74**, T662-T663

23. Garay, B.M. (1994b): Discretization and Morse-Smale dynamical systems on planar discs. Acta Math. Univ. Comenianae **63**, 25–38

24. Garay, B.M. (1996a): Various closeness concepts in numerical ODE's. Computers Math. Applic. **31**, 113–119

25. Garay, B.M. (1996b): On C^j-closeness between the solution flow and its numerical approximation. J. Difference Eq. Appl. **2**, 67–86

26. Garay, B.M. (1996c): On structural stability of ordinary differential equations with respect to discretization. Numer Math. **72**, 449–479

27. Garay, B.M. (1996d): The discretized flow on domains of attraction: a structural stability result. IMA J. Numer. Anal. (submitted)

28. Garay, B.M. (1996e): Normally hyperbolic invariant manifolds and discretization. (in preparation)

29. Garay, B.M., Hofbauer, J. (1996): Chain recurrence and discretization. Bull. Austral. Math. Soc. (in print)

30. Hairer, E., Norsett, S.P., Wanner, G. (1993): Solving Ordinary Differential Equations I, Nonstiff Problems (2nd ed.). Springer, Berlin Heidelberg New York

31. Hale, J.K. (1994): Numerical dynamics. Contemp. Math. **172**, 1-30

32. Hamilton, R.S. (1982): The inverse function theorem of Nash and Moser. Bull. Amer. Math. Soc. **7**, 65–222

33. Hassard, B., Zhang, J. (1994): Existence of a homoclinic orbit of the Lorenz system by precise shooting. SIAM. J. Math. Anal. **25**, 179-196

34. Hirsch, M.W. (1976): Differential Topology. Graduate Texts in Mathematics **33**, Springer, Berlin Heidelberg New York

35. Hirsch, M.W., Pugh, C., Shub, M. (1977): Invariant Manifolds. Lecture Notes in Mathematics **583**, Springer, Berlin Heidelberg New York

36. Irwin, M.C. (1980): Smooth Dynamical Systems. Academic Press, New York

37. Kloeden, P.E., Lorenz, J. (1986): Stable attracting sets in dynamical systems and their one-step discretizations. SIAM J. Num. Anal. **23**, 986–995

38. Lorenz, J. (1994): Numerics of invariant manifolds and attractors. Contemp. Math. **172**, 185-202

39. Milnor, J. (1984): Remarks on infinite-dimensional Lie groups. In: B.S. DeWitt, R. Stora eds. Relativité, groupes et topologie II, pp. 1007–1057. North-Holland, Amsterdam

40. Mischaikow, K., Mrozek, M. (1995): Chaos in the Lorenz equations: a computer-assisted proof. Bull. Amer. Math. Soc. **32**, 66–72

41. Mischaikow, K., Mrozek, M. (1996): Chaos in the Lorenz equations: a computer-assisted proof, Part II: details. (preprint)

42. Moore, G. (1995): Computation and parametrization of invariant curves and tori. SIAM J. Numer. Anal. (submitted)

43. Mrozek, M. (1990): Leray functor and cohomological Conley index for discrete time dynamical systems. Trans. Amer. Math. Soc. **318**, 149-178

44. Mrozek, M. (1995): The Conley index and rigorous numerics. (this volume)

45. Moser, J. (1969): On a theorem of Anosov. J. Diff. Eq. **5**, 411–440

46. Neumaier, A., Rage, T. (1993): Rigorous chaos verification in discrete dynamical systems. Physica D **67**, 327–346

47. Palmer, K. Chow, S.N. (1992): On the computation of orbits of dynamical systems: the higher dimensional case. J. Complexity **8**, 398–423

48. Phelps, R.R. (1989): Convex Functions, Monotone Operators and Differentiability. Springer, Berlin Heidelberg New York

49. Pollicott, M. (1993): Lectures on Ergodic Theory and Pesin Theory on Compact Manifolds. Cambridge University Press, Cambridge

50. Reinfelds, A. (1994): The reduction principle for discrete dynamical and semidynamical systems in metric spaces. J. Appl. Math. Phys. **45**, 933-955

51. Robbin, J.W. (1971): A structural stability theorem. Ann. of Math. **94**, 447–493

52. Robinson, C. (1974): Structural stability of vector fields. Ann. of Math. **99**, 154–175

53. Sauer, T., Yorke, J.A. (1991): Rigorous verification of trajectories for the computer simulation of dynamical systems. Nonlinearity **4**, 961–979

54. Shub, M. (1987): Global Stability of Dynamical Systems. Springer, Berlin Heidelberg New York

55. Srzednicki, R. (1995): On geometric detection of periodic solutions and chaos. (this volume)

56. Stuart, A.M. (1994): Numerical analysis of dynamical systems. In: A. Iserles ed., Acta Numerica, pp. 467–572. Cambridge University Press, Cambridge

57. Stuart, A.M., Humphries, A.R. (1996): Dynamical Systems and Numerical Analysis. Cambridge University Press, Cambridge

58. Wiggins, S. (1988): Global Bifurcations and Chaos: Analytical Methods. Springer, Berlin Heidelberg New York

59. Zgliczynski, P. (1995): A computer-assisted proof of the horseshoe dynamics-in the Henon map. Random Comput. Dynamics (submitted)

THE CONLEY INDEX AND RIGOROUS NUMERICS

M. Mrozek
Jagiellonian University, Cracow, Poland

1 Introduction

The Conley index is a topological invariant of isolated invariant sets, invented by C. Conley and his students in early 70's. The theory originated from the paper on existence of smooth isolating blocks by Conley and Easton [5]. The first concept of the index appeared in a conference announcement by Conley [3], in which main ideas of the homotopy and cohomological index were presented. The basic idea of the Conley index is to generalize the Morse index so as to make it defined also in the degenerate case and to give it features similar to the fixed point index (homotopy property, additivity property etc.).

The foundations of the theory, in a locally compact setting, were established in the papers of R. Churchill [2] and J. Montgomery [55] (the cohomological index), in the book of C. Conley [4] (the homotopy index and the generalized Morse index) and in the paper of H. Kurland [33] (the Morse index). The generalization of the theory to the case of a non locally compact space was given by K. Rybakowski in [78, 79, 80]. This admitted direct applications of the theory to partial differential equations and functional differential equations. The more recent developments of the theory can be found in [1, 16, 18, 19, 20, 21, 22, 31, 34, 35, 39, 40, 41, 43, 57, 58, 59, 60, 61, 62, 65, 66, 77, 88, 89, 94, 95].

The Conley index theory found numerous applications in the theory of ordinary and partial differential equations and in dynamical systems (see [6, 7, 8, 9, 10, 11, 12, 13, 14, 15, 23, 24, 28, 29, 30, 42, 44, 45, 46, 47, 48, 72, 73, 74, 75, 76, 81, 82, 83, 84, 85, 86]). In recent years the theory also showed its importance in the field of rigorous numerics. The goal of rigorous numerics is to perform numerical analysis in such a way that the results of computations provide proofs of interesting results. Examples of such an approach may be found in [17, 25, 26, 27, 32, 36, 37, 50, 51, 52, 53, 70, 71, 93].

*Research supported by KBN, grant 0449/P3/94/06

The aim of this review paper is to present how the Conley index theory may be fruitfully applied to rigorus numerics.

Throughout the paper the sets of reals, nonnegative reals, integers, nonnegative intergers, natural numbers and rationals are denoted by $\mathbb{R}$, $\mathbb{R}^+$, $\mathbb{Z}$, $\mathbb{Z}^+$, $\mathbb{N}$, $\mathbb{Q}$, respectively. For any set X the notation $\mathcal{P}(\mathcal{X})$ will stand for the family of all subsets of X. If X is a metric space with metric d and $A \subset X$, then we denote the boundary, the interior and the closure of A respectively by $\operatorname{bd} A$, $\operatorname{int} A$ and $\operatorname{cl} A$. If $x \in X$ and $r > 0$ then $B(x, r)$ will denote the closed ball of center x and radius r. Similarly, if $A \subset X$ then we put $B(A, r) := \{y \in X \mid d(y, A) \leq r\}$. The Alexander–Spanier cohomology of a pair of topological spaces (X, X_0) other a fixed ring R will be denoted by $H^*(H, X_0; R)$ or briefly by $H^*(X, X_0)$.

2   The Conley index for flows

The Conley index is assigned to isolated invariant sets of dynamical systems. Assume X is a locally compact metric space and $\pi : X \times \mathbb{R} \to X$ is a flow on X. For a compact $N \subset X$ we define the invariant part of N by

$$\operatorname{Inv} N := \{x \in N \mid \forall t \in \mathbb{R} \; \pi(x, t) \in N\}.$$

We say that S is invariant if $S = \operatorname{Inv} S$. We say that S is an isolated invariant set if there exists a compact set N such that $S = \operatorname{Inv} N$ and $S \subset \operatorname{int} N$. The set N is then called an isolating neighborhood.

There are several ways to define the Conley index, depending on the tools used in the construction (the cohomological, homotopy, shape Conley index). Thus it is convenient to give an axiomatic definition of the Conley index.

Assume $\mathcal{E}$ is a category and $\mathcal{C}$ is a certain class of pairs (S, π), where S is an isolated invariant set of the flow π.

By a Conley index on $\mathcal{C}$ we mean a map $C : \mathcal{C} \to \mathcal{E}$, which satisfies the following four properties:

1. **Ważewski property**: If $C(S, \pi)$ is not the zero object of $\mathcal{C}$ then S is non-empty.

2. **Homotopy property**: Assume $\pi_s, s \in [0, 1]$, is a family of dynamical systems continuously depending on s and $N \subset X$. If $(\operatorname{Inv} N, \pi_s) \in \mathcal{C}$ for all $s \in [0, 1]$ then $C(\operatorname{Inv} N, \pi_s)$ does not depend on s.

3. **Additivity property**: If $(S_1, \pi), (S_2, \pi) \in \mathcal{C}$ then $(S_1 \cup S_2, \pi) \in \mathcal{C}$ and the Conley index of $S_1 \cup S_2$ is the co-product or product of the indices of S_1 and S_2.

4. **Normalization property**: Assume X and π are smooth and x_0 is a hyperbolic stationary point of π. Then $(\{x_0\}, \pi) \in \mathcal{C}$ and $C(\{x_0\}, \pi)$ is non-zero.

The above axioms were chosen among the most often used properties of the Conley index for flows. It is not known to the author if the axioms are independant and if they uniquely characterize the Conley index.

The definition of the homotopy (cohomological) Conley index for flows rests on the concept of a Ważewski set. A compact set $W \subset X$ is called a Ważewski set if

$$W^- := \{x \in W \mid \exists \epsilon > 0 \ : \phi(x,t) \notin W \text{ for } 0 < t < \epsilon\}$$

is closed. In order to define the Conley index one first proves the following two theorems.

Theorem 2.1 *Every isolating neighborhood N contains a Ważewski set W such that Inv $N \subset W$.*

Theorem 2.2 *The homotopy type of the quotient space W/W^- (the Alexander-Spanier cohomology $H^*(W,W^-)$) does not depend on the index pair but only on the isolated invariant set S.*

This common value is taken as the homotopy (cohomological) Conley index.

Though the homotopy index is more general and easier to introduce (all necessary definitions from homotopy theory are elementary), the cohomological Conley index is easier to apply, because the cohomology is easier to compute than the homotopy type. Thus we will restrict our attention to the cohomology Conley index in the sequel. The cohomology Conley index takes the form of a graded modulus if the cohomology coefficients are in a ring or a vector space if the coefficients are in a field.

The definition of the Conley index for semidynamical systems is similar (see [78, 79, 80]) but conceptually more complicated, so we omit the details.

3 The Conley index for maps

The first step towards applying the Conley index in rigorous numerics is constructing it for discrete dynamical systems. This is because, when a differential equation is investigated numerically, a discrete numerical scheme approximating the equation is iterated. Recall that the discrete dynamical system on X is a group $\pi = \{\pi_t\}_{t \in \mathbb{Z}}$ of homeomorphisms of X. Unlike the flow, a discrete dynamical system is generated by a single element, namely π_1 (or π_{-1}). Hence it is often identified with its generator. In other words, if $f : X \to X$ is a homeomorphism then we can think of f as a discrete dynamical system $\pi_f = \{f^n\}_{n \in \mathbb{Z}}$ given by the iterates of f.

Already Ch. Conley [4] observed that the construction of an analog of the index for discrete dynamical systems would be of interest but the first construction (in smooth setting) is due to Robbin and Salamon [77]. Parts of the theory (the notions of isolating neighborhood and isolated invariant set) can be carried over directly to the discrete case. However, the notion of Ważewski set does not make sense in the discrete case.

This is because the trajectories may jump over the boundary. Thus it is necessary to extend the concept of exit set in such a way that it is not necessarily a subset of boundary. This is done via so called index pairs.

Definition 3.1 *The pair $P = (P_1, P_2)$ of compact subsets of N will be called an index pair of S in N iff the following three conditions are satisfied*

$$x \in P_i, \ f(x) \in N \ \Rightarrow \ f(x) \in P_i, \ \ i = 1, 2$$

$$x \in P_1, \ f(x) \notin N \ \Rightarrow \ x \in P_2$$

$$\text{Inv } N \subset \text{int}(P_1 \backslash P_2).$$

Like in the case of a flow (Theorem 2.1) one can prove

Theorem 3.2 *Every isolating neighbourhood admits an index pair P.*

However there is no direct analog of Theorem 2.2. The cohomology of index pair does depend on its choice. The proof fails, because the fundamental tool in the proof of Theorem 2.2, i.e. the homotopy built along trajectories of the flow, does not make any sense in the discrete case. In this respect entirely new ideas are needed. We follow here the author's approach presented in [58].

Let $P = (P_1, P_2)$ be an index pair in N. It can be easily derived from the definition of the index pair that f induces a map of pairs

$$f_P : (P_1, P_2) \ni x \rightarrow f(x) \in (P_1 \cup f(P_2), P_2 \cup f(P_2))$$

and the inclusion

$$i_P : (P_1, P_2) \ni x \rightarrow x \in (P_1 \cup (P_2), P_2 \cup f(P_2)).$$

induces an isomorphism in Alexander-Spanier cohomology.

Definition 3.3 *The endomorphism*

$$H^*(f_P) \circ H^*(i_P)^{-1} \ of \ H^*(P),$$

is called the (cohomological) index map associated with the index pair P and denoted by I_P.

The index map contains information which is essential in the construction of the discrete Conley index. This information is not important in the continuous case, because it is then trivial in the sense that the index map associated with the time-one translation map of a flow is always the identity.

In order to make use of the extra information we need some definitions. Together with each category $\mathcal{E}$ we consider the category of endomorphisms of $\mathcal{E}$ denoted by

Endo($\mathcal{E}$). The objects of Endo($\mathcal{E}$) are pairs (A, a), where $A \in \mathcal{E}$ and $a \in \mathcal{E}(A, A)$ is a distinguished endomorphism of A. The set of morphisms from $(A, a) \in \mathcal{E}$ to $(B, b) \in \mathcal{E}$ is the subset of $\mathcal{E}(A, B)$ consisting of exactly those morphisms $\varphi \in \mathcal{E}(A, B)$ for which $b\varphi = \varphi a$. We write $\varphi : (A, a) \to (B, b)$ to denote that φ is a morphism from (A, a) to (B, b) in Endo($\mathcal{E}$). We define the category of automorphisms of $\mathcal{E}$ as the full subcategory of Endo($\mathcal{E}$) consisting of pairs $(A, a) \in$ Endo($\mathcal{E}$) such that $a \in \mathcal{E}(A, A)$ is an automorphism, i.e., both an endomorphism and an isomorphism in $\mathcal{E}$. The category of automorphisms of $\mathcal{E}$ will be denoted by Auto($\mathcal{E}$). There is a functorial embedding

$$\mathcal{E} \ni A \to (A, \underset{A}{\mathrm{id}} \in \mathrm{Auto}(\mathcal{E})),$$

$$\mathcal{E}(A, B) \ni \varphi \to \varphi \in \mathrm{Auto}(\mathcal{E})(A, B),$$

hence we can consider the category $\mathcal{E}$ as a subcategory of Auto($\mathcal{E}$). Assume $\mathcal{C}$ is a full subcategory of Endo($\mathcal{E}$) and $L : \mathcal{C} \to$ Auto($\mathcal{E}$) is a functor. Let $(A, a) \in \mathcal{C}$. Then $L(A, a)$ is an object of Auto($\mathcal{E}$). Let a' denote the automorphism distinguished in $L(A, a)$. Obviously $a : (A, a) \to (A, a)$ is a morphism in Endo($\mathcal{E}$) and since $\mathcal{C}$ is a full subcategory of Endo($\mathcal{E}$) it is also a morphism in $\mathcal{C}$. Hence $L(a)$ is defined and it is a morphism from $L(A, a)$ to $L(A, a)$ in Auto($\mathcal{E}$). However, it need not be $L(a) = a'$ in general. $L(a)$ need not be even an isomorphism in Auto($\mathcal{E}$). We say that $L : \mathcal{C} \to$ Auto($\mathcal{E}$) is normal, if for each $(A, a) \in \mathcal{C}$ the morphism $L(a)$ is equal to the automorphism distinguished in $L(A, a)$.

There are many ways to define normal functors and actually one can prove the existence of a universal normal functor (see Szymczak [91]). One of the simplest constructions applies to the category of finitely dimensional graded vector spaces $\mathcal{V}$.

Let $(F, f) \in$ Endo($\mathcal{V}$). Define the generalized kernel of f as

$$\mathrm{gker}(f) := \bigcup\{f^{-n}(0)|n \in \mathbf{N}\}.$$

Since $f(\mathrm{gker}(f)) \subset \mathrm{gker}(f)$, we have an induced monomorphism

$$f' : F/\mathrm{gker}(f) \ni [x] \to [f(x)] \in F/\mathrm{gker}(f).$$

Put

$$L(F, f) := (F/\mathrm{gker}(f), f').$$

It is easy to verify that f' is in fact an isomorphism. Thus $L(F, f) \in \mathcal{EI}$. One can easily extend the above definition to morphisms, so that we indeed obtain a covariant functor

$$L : \mathcal{EE} \to \mathcal{EI}.$$

We will call it the Leray functor. One easily verifies that the Leray functor is normal. Other examples of normal functors may be obtained in terms of the direct and inverse limits (see [62]).

Let $\mathcal{M}$ denote the category of graded moduli over the ring R and let $L :$ Endo($\mathcal{M}$) $\to$ Auto($\mathcal{M}$) be a normal functor. The analog of Theorem 2.2 in the discrete case is the following theorem.

Theorem 3.4 *Assume $f : X \to X$ is a homeomorphism and S is an isolated invariant set with respect to π_f. Then $L(H^*(P), I_P)$ does not depend on the index pair but only on the isolated invariant set S.*

The above theorem allows us to define the Conley index in the discrete case as $L(H^*(P), I_P)$. Let us emphasize that in the discrete case the Conley index has the form of a pair

$$\mathrm{Con}^*(S) = (CH^*(S), \chi^*(S)),$$

where $CH^*(S)$ is a graded vector space and $\chi^*(S) : CH^*(S) \to CH^*(S)$ is a graded automorphism.

As we stated above, if f is the time-one-map of a flow and S is an isolated invariant set, then I_P is an identity. Consequently

$$\mathrm{Con}^*(S, \pi_f) = L(H^*(P), \mathrm{id}) = (H^*(P), \mathrm{id}) = H^*(P).$$

Hence we have the following

Theorem 3.5 *(sec [61])* *The cohomological Conley index of an isolated invariant set of a flow coincides with the corresponding cohomological Conley index of the time-one-map of this flow.*

There is also

Theorem 3.6 *(see [61]).* *S is an isolated invariant set with respect to a flow iff it is an isolated invariant set with respect to the time-one translation of the flow.*

Hence our index can be considered as a generalization of the cohomological Conley index for flows.

4 A horseshoe example

In this section we want to show how the Conley index for maps may be computed directly from its definition. For the convenience of the reader we first briefly recall basic definitions and notation concerning the Alexander–Spanier cohomology.

For a locally compact metric space X and $n \in \mathbf{N}$, let $\Gamma^n(X)$ denote the $\mathbf{Z}$-module of all functions $\varphi : X^{n+1} \to \mathbf{Z}$. The coboundary homomorphism $\delta : \Gamma^n(X) \to \Gamma^{n+1}(X)$ is defined by the formula

$$\delta\varphi(x_0, \ldots, x_{n+1}) = \sum_{i=0}^{n+1} (-1)^i \varphi(x_0, \ldots, x_{i-1}, x_{i+1}, \ldots, x_{n+1}).$$

Since $\delta\delta = 0$, $\Gamma^*(X) := (\Gamma^n(X), \delta)$ is a cochain complex over $\mathbf{Z}$. For $\varphi \in \Gamma^n(X)$ define the support of φ by $|\varphi| :=$

$$\{x \in X \mid \forall\, V \subset X \text{ open } x \in V \Rightarrow \exists\, x_0, \ldots, x_n \in V \text{ s.t. } \varphi(x_0, \ldots, x_n) \neq 0\}.$$

Then, $\Gamma_0^n(X) := \{\varphi \in \Gamma^n(X) \mid |\varphi| = \emptyset\}$ is a submodule of $\Gamma^n(X)$ and $\Gamma_0^*(X) := (\Gamma_0^n(X), \delta)$ is a cochain subcomplex of $\Gamma^*(X)$. The quotient cochain complex $\Gamma^*(X)/\Gamma_0^*(X)$ is denoted by $\bar{\Gamma}^*(X)$. If $u \in \bar{\Gamma}^*(X)$ then $|\varphi|$ does not depend on φ for all representations $\varphi \in u$. Hence one can set $|u| := |\varphi|$, where $\varphi \in u$.

If Y is another topological space and $g : X \to Y$ is continuous then there is an induced cochain map $g^\# : \Gamma^*(Y) \to \Gamma^*(X)$ defined by the formula

$$(g^\# \varphi)(x_0, \dots, x_n) = \varphi(g(x_0), \dots, g(x_n)).$$

Since $g^\#(\Gamma_0^*(Y)) \subset \Gamma_0^*(X)$, there is also an induced map $g^\# : \bar{\Gamma}^*(Y) \to \bar{\Gamma}^*(X)$.

If $i : X_0 \hookrightarrow X$ is an inclusion, then $\bar{\Gamma}^*(X, X_0) := \ker i^\#$ is a cochain subcomplex of $\bar{\Gamma}^*(X)$. The Alexander–Spanier cohomology of the pair (X, X_0) is defined as the homology of the cochain subcomplex $\bar{\Gamma}^*(X, X_0)$ and denoted by $\bar{H}^*(X, X_0)$. If (Y, Y_0) is another topological pair and $\kappa : \bar{\Gamma}^*(Y, Y_0) \to \bar{\Gamma}^*(X, X_0)$ is a cochain map then it induces the map $\kappa^* : \bar{H}^*(Y, Y_0) \to \bar{H}^*(X, X_0)$. In particular if $g : (X, X_0) \to (Y, Y_0)$ is a continuous map then one has a map $(g^\#)^* : \bar{H}^*(Y, Y_0) \to \bar{H}^*(X, X_0)$, which for the sake of simplicity will be denoted by g^*.

Proposition 4.1 (see [38], Sect. 1.1) *Assume $\varphi \in \Gamma^*(X)$. Then*

$$[\varphi] \neq 0 \quad \Rightarrow \quad |\varphi| \neq \emptyset.$$

Proposition 4.2 .see [50], Prop. 3.5 *Assume $X_0, X_1 \subset X$ are closed subsets of X such that*

$$X = \operatorname{int} X_0 \cup \operatorname{int} X_1.$$

Then

$$\bar{H}^*(X, X_0 \cap X_1) \cong \bar{H}^*(X, X_0) \oplus \bar{H}^*(X, X_1)$$

One can show (see [56]) that every isolating neighborhood admits a regular index pair, i.e. an index pair P which additionally satisfies the following condition

$$\operatorname{cl}(P_1 \backslash P_2) \cap \operatorname{cl}(f(P_2) \backslash P_1) = \emptyset,$$

In case of regular index pairs one can define the index map also on the chain level. For this end we need the following proposition

Proposition 4.3 . *If P is a regular index pair then $i_P : (P_1, P_2) \to (P_1 \cup f(P_2), P_2 \cup f(P_2))$ induces an isomorphism $i^\# : \bar{\Gamma}^*(P_1 \cup f(P_2), P_2 \cup f(P_2)) \to \bar{\Gamma}^*(P_1, P_2)$ and*

$$(i_P^{\#,P})^{-1}([\varphi]) = [\lambda_P(\varphi)]$$

where $\lambda_P : \Gamma^*(P_1, P_2) \to \Gamma^*(P_1 \cup f(P_2), P_2 \cup f(P_2))$ is defined by

$$\lambda_P(\varphi)(x_0, \dots, x_p) = \begin{cases} \varphi(x_0, \dots, x_p) & \text{if } x_0, \dots, x_p \in P_1 \\ 0 & \text{otherwise.} \end{cases}$$

Proof: The proposition follows easily from the proof of Lemma 4 in [90], Sect. 6.4. QED

The cochain map $f^\# \circ (i^\#)^{-1}$ is called the cochain index map and is denoted by $f_P^\#$. Obviously

$$f_P^* = (f_P^\#)^*.$$

Let $J_0 := [-12, -6]$, $J_1 := [6, 12]$, $J := [-12, 12]$. For $i = 0, 1$ define the maps

$$\mu_i : J_i \ni s \to 6s + (-1)^i 54 \in \mathbb{R}$$

$$\nu_i : [-18, 18] \ni s \to (s - (-1)^i 54)/6 \in J_i$$

Put $N := J \times J$, $N_i := J \times J_i$ and for $i = 0, 1$ define maps

$$f_i : N_i \ni (x, y) \to (\mu_i(x), \nu_i(y)) \in \mathbb{R}^2.$$

Now extend the map $f_0 \cup f_1$ to a homeomorphism $f : \mathbb{R}^2 \to \mathbb{R}^2$ in such a way that $f(N \setminus (N_0 \cup N_1)) \cap N = \emptyset$. An elementary computation shows that

$$f^{-1}(N) \cap N \cap f(N) = \bigcup_{i,j=0,1} K_i \times K_j,$$

where $K_0 := [-11, -7]$, $K_1 := [7, 11]$. In particular we obtain

$$\text{Inv}(N, f) \subset f^{-1}(N) \cap N \cap f(N) \subset \text{int } N,$$

which shows that N is an isolating neighborhood for f.

One easily verifies that (N, E), where $E := [-12, -11] \cup [-7, 7] \cup [11, 12]$ is an index pair for f in N. Taking $E_0 := [-12, -11] \cup [-7, 12]$ and $E_1 := [-12, 7] \cup [11, 12]$ we get $E = E_0 \cap E_1$ and by Proposition 4.2 we obtain

$$H^*(N, E) = H^*(N, E_0) \oplus H^*(N, E_1).$$

Using the exact sequence of a pair (see [38], Sect.8.2) we conclude that

$$H^i(N, E_0) = H^i(N, E_1) = \begin{cases} \mathbb{Z} & \text{for } i = 1, \\ 0 & \text{otherwise.} \end{cases}$$

and

$$H^i(N, E) = \begin{cases} \mathbb{Z}^2 & \text{for } i = 1, \\ 0 & \text{otherwise.} \end{cases}$$

Let us turn now our attention to computing the index map. For $s \in J$ define the map

$$\varphi_s : N^2 \ni ((x_0, y_0), (x_1, y_1)) \to \begin{cases} 1 & \text{if } x_0 < s, x_1 \geq s, \\ -1 & \text{if } x_0 \geq s, x_1 < s, \\ 0 & \text{otherwise.} \end{cases}$$

and let $u_s := \varphi_s + \Gamma_0^1(N)$. One easily finds that $|u_s| = \{(x,s) \mid x \in J\}$. In particular we get

$$u_s \in \left\{ \begin{array}{ll} \bar{\Gamma}^1(N, E_0) & \text{for } s \in L_0 \\ \bar{\Gamma}^1(N, E_1) & \text{for } s \in L_1. \end{array} \right.$$

where $L_0 := (-11, -7)$, $L_1 := (7, 11)$. Let $s_1, s_2 \in L_0$, $s_1 < s_2$. Define

$$\psi_{s_1, s_2} : N \ni (x, y) \to \left\{ \begin{array}{ll} 1 & \text{if } s_2 > x \geq s_1, \\ 0 & \text{otherwise.} \end{array} \right.$$

and put $v_{s_1, s_2} := \psi_{s_1, s_2} + \Gamma_0^0(N)$. An elementary computation shows that $\delta v = u_{s_1} - u_{s_2}$ and $\delta u_s = 0$, i.e. u_{s_1}, u_{s_2} are homologous co-cycles in $\bar{\Gamma}^1(N, E_0)$. An analogous argument shows that u_{s_1}, u_{s_2} are homologous co-cycles in $\bar{\Gamma}^1(N, E_1)$ for $s_1, s_2 \in L_1$, $s_1 < s_2$. Let $\alpha_i := [u_s]$, where $s \in L_i$.

Since by Proposition 4.1 $u_s \neq 0$ and obviously u_s is not a non-trivial multiple of another function, we conclude that α_i generates $H^1(N, E_i)$ for $i = 0, 1$.

Now we easily verify that

$$f_P^*(\alpha_0) = [f_P^\#(u_{-9})] = [u_{7.5} + u_{-10.5}] = [u_{7.5}] + [u_{-10.5}] = [u_9] + [u_{-9}] = \alpha_0 + \alpha_1$$

and

$$f_P^*(\alpha_1) = [f_P^\#(u_9)] = [u_{-7.5} + u_{10.5}] = [u_{-7.5}] + [u_{10.5}] = [u_{-9}] + [u_9] = \alpha_0 + \alpha_1.$$

Thus the matrix of f_P^* in the basis α_0, α_1 is

$$A := \left[\begin{array}{cc} 1 & 1 \\ 1 & 1 \end{array} \right]$$

Since $A^n = 2^{n-1} A$, we see that gker $A = \ker A$ and we easily conclude that in case of coefficients in the field of rationals and the Leray functor

$$\mathrm{Con}^n(N, f) = \left\{ \begin{array}{ll} (\mathbb{Q}, 2\,\mathrm{id}) & \text{if n=1,} \\ 0 & \text{otherwise.} \end{array} \right.$$

5   Properties of the Conley index for maps

Similarly to the Conley index for flows the Conley index for maps has several properties which faciliate its computation. Some of them are analogous to the properites of the Conley index for flows but some are specific to the discrete case. In this section we would like to summarize them.

Theorem 5.1 Ważewski property: *If N is an isolating neighbourhood with respect to π_f and $\mathrm{Con}(\mathrm{Inv}\,N, \pi_f) \neq 0$ then $\mathrm{Inv}\,N$ is non-empty.*

Theorem 5.2 Stability property: *If N is an isolating neighbourhood with respect to π_f then it is an isolating neighbourhood with respect to π_g for g sufficiently close to f in the compact-open topology and the Conley indexes with respect to f and g coincide.*

Theorem 5.3 Homotopy property: *Assume π_s, $s \in [0,1]$, is a family of dynamical systems on X continuously depending on s. If $N \subset X$ is an isolating neighbourhood with respect to every dynamical system π_s then $\mathrm{Con}(\mathrm{Inv}\, N, \pi_s)$ does not depend on s.*

Theorem 5.4 Additivity property: *If N, N_1, N_2 are isoltaing neighbourhoods such that $\mathrm{Inv}\, N = \mathrm{Inv}\, N_1 \cup \mathrm{Inv}\, N_2$, $\mathrm{Inv}\, N_1 \cap \mathrm{Inv}\, N_2 = \emptyset$ then*

$$\mathrm{Con}(\mathrm{Inv}\, N, \pi_f) = \mathrm{Con}(\mathrm{Inv}\, N_1, \pi_f) \times \mathrm{Con}(\mathrm{Inv}\, N_2, \pi_f)$$

Theorem 5.5 Normalization property: *The whole space X is an isolating neighbourhood and $\mathrm{Con}(X, \pi_f) = (H^*(X), f^*)$.*

Theorem 5.6 Commutativity property *(see [61]). Assume $f = \psi\phi, g = \phi\psi$, where $\phi : X \to Y, \psi : Y \to X$ are continuous,. If $S \subset X$ is an isolated invariant set with respect to π_f then $\phi(S)$ is an isolated invariant set with respect to π_g and $\mathrm{Con}(S, \pi_f) = \mathrm{Con}(\phi(S), \pi_g)$.*

Theorem 5.7 Relation to fixed point index *(see [58]). Assume X is a compact ANR and S is an isolated invariant set of π_f. Let $\mathrm{Con}(S, \pi_f) = (E, e) \in \mathcal{EI}$. Then E is of finite type and $\Lambda(e)$, the Lefschetz number of e is exactly the fixed point index of f in a neighbourhood of S.*

Let x_0 be a hyperbolic fixed point of f. Let k denote the number of eigenvalues of $\mathrm{Df}(x_0)$ with modulus greater than one (counted with multiplicity). Let l denote the number of real eigenvalues of $\mathrm{Df}(x_0)$ which are less than -1. Then the pair (k, l) will be called the Morse index of x_0.

Theorem 5.8 *Assume x_0 is a hyperbolic fixed point of a C^1-diffeomorphism $f : \mathbb{R}^n \to \mathbb{R}^n$. Then $\{x_0\}$ is an isolated invariant set and*

$$\mathop{\mathrm{Con}}_{i}(\{x_0\}, \pi_f) = \begin{cases} 0 & \text{for } i \neq k \\ (\mathbb{Q}, (-1)^l \,\mathrm{id}) & \text{for } i = k, \end{cases}$$

where (k,l) is the Morse index of $\{x_0\}$.

6 The Conley index for multivalued discrete dynamical systems

The only way to obtain some rigorous information from numerical computations is to perform rigorous error analysis. Such an analysis is possible (see [63]). As an outcome we obtain a multivalued map: the exact values are not known but a set enclosing the exact values is computer rigorously. This makes necessary extending the Conley index theory to multivalued maps. Such a theory for multivalued flows was proposed in [57]. The generalization in that case is quite natural. However, the discrete case is different, because it is not evident how to define an isolating neighbourhood in that case.

Let us recall that a *multivalued (mv) map* is a map $F : \mathbb{R}^p \to \mathcal{P}\mathbb{R}^{\mathrm{II}}$. For $A \subset \mathbb{R}^p, B \subset \mathbb{R}^q$ the *image of A, weak preimage of B and strong preimage of B* are defined by

$$F(A) := \bigcup \{ F(x) \mid x \in A \}$$

$$F^{*-1}(B) := \{ x \in X \mid F(x) \cap B \neq \emptyset \}$$

$$F^{-1}(B) := \{ x \in \operatorname{dom} F \mid F(x) \subset B \}$$

F is *upper semicontinuous (usc)* if $F^{-1}(U)$ is open for any open $U \subset \mathbb{R}^q$.

In [31] an extension of the Conley index theory, based on the following definition of the isolating neighbourhood, was proposed.

Definition 6.1 *Assume $F : X \to \mathcal{P}X$ is a multivalued map and $N \subset X$. The set* $\operatorname{Inv}(N, F)$ *is defined as the set of $x \in N$ such that there exists a function $\sigma : \mathbf{Z} \to N$ satisfying $\sigma(0) = x$ and $\sigma(n + 1) \in F(\sigma(n))$. The set N is said to be an isolating neighbourhood for F if*

$$B_{\operatorname{diam}_N F}(\operatorname{Inv} N) \subset \operatorname{int} N,$$

where $\operatorname{diam}_N F$ is the maximal diameter of the values of F in N.

The generalization of the notion of the index pair is then straightforward:

Definition 6.2 *Let N be an isolating neighbourhood for F. A pair $P = (P_1, P_2)$ of compact subsets $P_2 \subset P_1 \subset N$ is called an* index pair *if the following conditions are satisfied:*

$$F(P_i) \cap N \subset P_i, \ i = 1, 2;$$

$$F(P_1 \backslash P_2) \subset N;$$

$$\operatorname{Inv} N \subset \operatorname{int}(P_1 \backslash P_2)$$

The definition of the index map does not essentially differ from the single valued case. One only needs to put some admissibility conditions to ensure that the multivalued map induces a map on cohomology level. This is all what is necessary to develop the Conley index theory for multivalued discrete dynamical systems.

Obviously not all multivalued maps may be treated by computer. To do this some finite way of representing these maps is necessary. For this end let us introduce the following definitions.

Let $\hat{\mathbb{R}} \subset \mathbb{R}$ be a given finite set. *Elementary representable sets (over $\hat{\mathbb{R}}$)* are the elements of

$$\mathcal{E} := \{A_1 \times A_2 \times \cdots \times A_d \mid$$

$$A_i = \{a\} \text{ or } A_i = (a, b) \text{ where } a, b \in \hat{\mathbb{R}}, \ a << b\},$$

where $a << b$ means that a is an immediate predecesor of b.

Representable sets are the elements of

$$\mathcal{R} := \{E_1 \cup E_2 \cup \ldots \cup E_k \mid E_i \in \mathcal{E}\}$$

A mv map $F : \mathbb{R}^p \to \mathcal{P}\mathbb{R}^{\amalg}$ is *representable* if

$$\text{dom } F := \{x \in \mathbb{R}^p \mid F(x) \neq \emptyset\} \in \mathcal{R},$$

$$\forall x \in \text{dom } F \ F(x) \in \mathcal{R},$$

$$\forall E \in \mathcal{E} \ F_{|E} = \text{const},$$

If $U \subset \mathbb{R}^p$ and $f : U \to \mathbb{R}^q$ then $F : \mathbb{R}^p \to \mathcal{P}\mathbb{R}^{\amalg}$ is a *representation of f* if F is representable and f is a *selector* of $F_{|U}$, i.e.

$$U \subset \text{dom } F \text{ and } \forall x \in U \ f(x) \in F(x).$$

Assume that $\{\hat{\mathbb{R}}_n\}$ is a given, monotonically increasing sequence of finite subsets of $\mathbb{R}$ such that $\bigcup \{\hat{\mathbb{R}}_n\}$ is dense in $\mathbb{R}$. We have the following theorem.

Theorem 6.3 *(see [31, 63]) If f is L-Lipschitz continuous, then there exists a sequence $\{F_n\} : \mathbb{R}^p \to \mathcal{P}\mathbb{R}^{\amalg}$ of usc multivalued representations of f such that $F_n \to f$.*

The above theorem shows that, given a sufficiently powerful computer, any Lipschitz continuous map admits an arbitrarily close multivalued representation.

7 Inheritable properties

Multivalued representations of single valued maps enable passing from the world of continua to the world of finite mathematics. To be able to go back one needs the concept of inheritance. Assume $\mathcal{A}$ is a collection of mv maps and $\varphi(F)$ is a property of such maps. We say that φ is *inheritable* if for every single-valued selector f of F in $\mathcal{A}$

$$\varphi(F) \ \Rightarrow \ \varphi(f)$$

We say that an inheritable property φ is *strongly inheritable* if for any single valued map $f \in \mathcal{A}$ such that $\varphi(f)$ and for any sequence $\{F_n\} \subset \mathcal{A}$ satisfying $F_n \to f$ we have $\varphi(F_n)$ for n sufficiently large.

If $\alpha(F)$ is a term then we say that α is inheritable (strongly inheritable) if for any x the property $\alpha(F) = x$ is inheritable (strongly inheritable).

Example 7.1 *Take $\mathcal{A} = \{F : \mathbb{R}^p \to \mathcal{P}\mathbb{R}\}$ and consider the following three properties*

$$\alpha(F) \Leftrightarrow \exists x \in \mathbb{R}^p \ F(x) > 0$$

$$\beta(F) \Leftrightarrow \exists x \in \mathbb{R}^p \ F(x) \geq 0$$

$$\gamma(F) \Leftrightarrow \exists x \in \mathbb{R}^p \ F(x) \ni 0$$

The first property is strongly inheritable. The second property is inheritable but not strongly inheritable. The last property is not inheritable.

Example 7.2 *Assume $\mathcal{A} = \{F : \mathbb{R}^p \to \mathcal{P}\mathbb{R}\sqrt{}\}$ and let B be the unit ball in $\mathbb{R}^p$. Consider the following three properties*

$$\mu(F) \Leftrightarrow F(B) \subset \text{int } B$$

$$\nu(F) \Leftrightarrow F(B) \subset B$$

$$\kappa(F) \Leftrightarrow \exists x \in B \ x \in F(x)$$

Again the first property is strongly inheritable, the second property is inheritable but not strongly inheritable and the last property is not inheritable.

Remark 7.3   *1. Multivalued representable maps behave like combinatorial objects; operations on them are algorithmizable.*

   2. Multivalued representable usc maps are conceptually close to ordinary continuous maps, so some theories carry over from single-valued continuous maps to multivalued usc maps. This allows for many natural inheritable properties.

   3. Multivalued representations of single valued maps may be used to prove inheritable properties of single valued maps.

   4. If a single valued map satisfies a strongly inheritable property then a sufficiently powerful computer will prove it.

   5. Usually properties of real interest are non-inheritable

   6. To get interesting applications we need theorems which convert inheritable properties to non-inheritable properties

An example of such a theorem, which converts property α in Example 7.1 to property γ, is the Darboux Theorem. Similarly, the theorem, which converts property μ in Example 7.2 to property κ is the Brouwer Fixed Point Theorem.

The following theorem is essential in applications of the Conley index to rigorous numerics.

Theorem 7.4 *(see [63]). The isolating neighborhood, the index pair, and the Conley index are strongly inheritable terms.*

Of course, in order to be able to compute the Conley index of a multivalued representation we need to know if there exists an index pair consisting of representable sets and how to construct it.

This is relatively easy under extra assumptions as in the following theorem.

Theorem 7.5 *(see [63]) If $F : \mathbb{R}^d \to \mathcal{P}\mathbb{R}^\Gamma$ is usc representable and N is a representable set such that*

$$B_{\mathrm{diam}_N F}(F^{*-1}(N) \cap N \cap F(N)) \subset \mathrm{int}\, N.$$

then N is an isolating neighborhood and $(N, N\backslash F^{-1}(\mathrm{int}\, N))$ is an index pair in N consisting of representable sets.

A general answer to the question of the existence and construction of representable index pairs may be found in [64].

8   Chaos in the Lorenz equations

The world of representable multivalued maps and inheritable properties opens the way to rigorous computations of the Conley index for concrete dynamical systems. Since there are various qualitative descriptions of dynamics in terms of the Conley index, many new interesting results may be obtained this way. As an example let us announce a result concerning chaotic dynamics in the classical case of the Lorenz equations (see [53] and also [51, 52]). The result is based on a generalization of the Conley index theory proposed by Szymczak [92].

For a $k \times k$ matrix $A = (A_{ij})$ over $\mathbb{Z}_2$ put

$$\Sigma(A) := \{\alpha : \mathbb{Z} \to \{1, 2, \ldots, k\} \mid \forall i \in \mathbb{Z} A_{\alpha(i)\alpha(i+1)} = 1\}$$

$$\sigma : \Sigma(A) \ni \alpha \to (\mathbb{Z} \ni i \to \alpha(i+1) \in \{1, 2, \ldots k\} \in \Sigma(A).$$

Theorem 8.1 *(see [53]). Consider the Lorenz equations*

$$
\begin{aligned}
\dot{x} &= s(y - x) \\
\dot{y} &= Rx - y - xz \\
\dot{z} &= xy - qz,
\end{aligned}
\tag{1}
$$

and the plane $P := \{(x, y, z) \mid z = 27\}$. For all parameter values in a sufficiently small neighborhood of $(s, R, q) = (10, 28, 8/3)$ there exists a Poincaré section $N \subset P$ such that the associated Poincaré map is Lipschitz and well defined. Furthermore, for

$$A := \begin{bmatrix} 0 & 1 & 1 & 0 & 0 & 0 \\ 0 & 0 & 0 & 1 & 1 & 0 \\ 0 & 0 & 0 & 0 & 0 & 1 \\ 1 & 0 & 0 & 0 & 0 & 0 \\ 0 & 1 & 1 & 0 & 0 & 0 \\ 0 & 0 & 0 & 1 & 1 & 0 \end{bmatrix}$$

there is a continuous surjection $\rho : \mathrm{Inv}(N, g) \to \Sigma(A)$ such that

$$\rho \circ g = \sigma \circ \rho.$$

Moreover, for every $\alpha \in \Sigma(A)$ which is periodic there exists an $x \in \mathrm{Inv}(N, g)$ on a periodic trajectory such that $\rho(x) = \alpha$.

The main tool in the proof of the above theorem is a criterion of chaos (see [53]) based on the Conley index theory. In order to verify the assumptions of that theorem we choose a candidate for a representable isolating neighborhood N for the Poincare map f and compute a multivalued representation F of f. If N turns out to be an isolating neighborhood for F, we compute an index pair and find the Conley index for F. By inheritance, f has the same Conley index in N. If the computed index satisfies the assumptions of the chaos criterion, which indeed is the case, the theorem is proved. If not, the proof fails. In case N fails to be an isolating neighborhood for F, one is free to find a better mv representation of f and repreat the verification process.

References

[1] V. Benci, A Generalization of the Conley-Index Theory, *Rendiconti Istituto Matematico di Trieste*, 18(1986),16-39.

[2] R. Churchill, Isolated Invariant Sets in Compact Metric Spaces, *J. Diff. Equ.* **12** (1972), 330-352.

[3] C. Conley, On a Generalization of the Morse Index, Ordinary Differential Equations, *1971 NRL-MRC Conference*, Ed. L.Weiss. Academic Press, New York (1972), 133-146.

[4] C. Conley, On a generalization of the Morse index, in *Ordinary Differential Equations*, 1971 NRL-MRC Conference, ed. L. Weiss, Academic Press, New York (1972), 27-33.

[5] C.Conley, R.Easton, Isolated Invariant Sets and Isolating Blocks, *Trans. AMS* **158** (1971), 35-61.

[6] C. Conley, R. Gardner, An Application of the Generalized Morse Index to Travelling Wave Solutions of a Competitive Reaction-Diffusion Model , *Indiana Univ. Math. Journ. 33(1984)*, 319-343.

[7] C. Conley, J. Smoller, Shock Waves as Limits of Progressive Wave Solutions of Higher-Order Equations I,II, *Comm. Pure Appl. Math.* 24(1971), 459-472, 25(1972), 131-146.

[8] C. Conley, J. Smoller, On the Structure of Magneto-Hydrodynamic Shock Waves, *Comm. Pure Appl. Math.* 28(1974), 367-375.

[9] C. Conley, J. Smoller, On the Structure of Magneto-Hydrodynamic Shock Waves II, *J. Math Pures et Appl.* 54(1975), 429-444.

[10] C. Conley, J. Smoller, Isolated Invariant Sets of Parameterized Systems of Differential Equations in: *The Structure of Attractors in Dynamical Systems* (N.G. Markley, J. C. Martin, W. Perizzo, Eds.), Lect. Notes in Math. 668, Springer Verlag, Berlin, 1978.

[11] C. Conley, J. Smoller, *Topological Techniques in Reaction-Diffusion Equations*, Springer Lecture Notes in Biomath. 38(1980), 473-483.

[12] C. Conley, J. Smoller, Bifurcations and Stability of Stationary Solutions of the Fritz-Hugh-Nagumo Equations, *J. Diff. Equ.* 63(1986), 389-405.

[13] C. Conley, J. Smoller, Remarks on Travelling Wave Solutions of Non-Linear Diffusion Equations, in: Lect. Notes in Math. 525 (ed. P. Hilton) (1975), 77-89.

[14] C. Conley, E. Zehnder, The Birkhoff-Lewis Fixed Point Theorem and a Conjecture by V.I. Arnold, *Inv. Math.* 73(1983), 33-49.

[15] C. Conley, E. Zehnder, Morse Type Index Theory for Flows and Periodic Solutions for Hamiltonian Equations, *Comm. on Pure and Appl. Math.* 37(1984), 207-253.

[16] M. Degiovanni, M. Mrozek, The Conley index for maps in absence of compactness, *Proc. Royal Soc. Edinburgh* 123A(1993),75-94.

[17] J.P Eckmann, H. Koch, P. Wittwer, A computer-assisted proof of universality for area-preserving maps, *Memoirs of the American Mathematical Society* **47**, 1984, 1-121.

[18] A. Floer, A Refinement of the Conley Index and an Application to the Stability of Hyperbolic Invariant Sets, *Ergod. Th. Dyn. Syst.* 7 (1987), 93-103.

[19] A.Floer, E.Zehnder, The Equivariant Conley Index and Bifurcations of Periodic Solutions of Hamiltonian Systems, *Ergod. Th. Dyn. Sys.* 8*(1988), 87-97.

[20] R. Franzosa, Index Filtrations and the Homology Index Braid for Partially Ordered Morse Decomposition, *Trans. AMS 298.1(1986)*, 193-213.

[21] R. Franzosa, The Continuation Theory for Morse Decompositions and Connection Matrices, *Trans. AMS.* 310.2(1988), 781-803.

[22] R. Franzosa, The Connection Matrix Theory for Morse Decompositions, *Trans. AMS* 311.2(1989), 561-592.

[23] R. Gardner, Existence of Multidimensional Travelling Wave Solutions of an Initial-Boundary Value Problem, *J. Diff. Equ.* 61(1986), 335-379.

[24] R. Gardner, J. Smoller, The Existence of Periodic Travelling Waves for Singularly Perturbed Predator-Prey Equations via the Conley Index, *J. Diff. Equ.* 47(1983), 133-161.

[25] B. Hassard and J. Zhang, Existence of a homoclinic orbit of the Lorenz system by precise shooting, *SIAM J. Math. Anal.*, **25**, 1994, 179-196.

[26] B. Hassard, S.P. Hastings, W.C. Troy, J. Zhang, A computer proof that the Lorenz equations have "chaotic" solutions, *Appl. Math. Letter*, to appear.

[27] S.P. Hastings, W.C. Troy, A shooting approach to the Lorenz equations, *Bull. AMS (N.S.)* **27**, 1992, 298-303.

[28] H.Hattori, K.Mischaikow, A Dynamical System Approach to a Phase Transition Problem, *J. Diff. Equ.,* to appear.

[29] V. Hutson, K. Mischaikow, Travelling Waves for Competing Species, in preparation.

[30] V. Hutson, K. Mischaikow, Travelling Waves for Competitive and Mixed Systems, in preparation.

[31] T. Kaczyński and M. Mrozek, Conley index for discrete multivalued dynamical systems, *Topology & its Appl*, accepted.

[32] H. Koch, A. Schenkel, P. Wittwer, Computer assisted proofs in analysis and programming in logic: a case study, Universite de Geneve, preprint.

[33] H.L. Kurland, The Morse Index of an Isolated Invariant Set is a Connected Simple System, *J. Diff. Equ.* **42** (1981), 234-259.

[34] H.L. Kurland, Homotopy Invariants of a Repeller-Attractor Pair: I. The Puppe Sequence of an R-A Pair, J. Diff. Equ. 46(1982), 1-31, II. Continuation of R-A pairs, *J. Diff. Equ.* 49(1983), 281-329.

[35] H.L. Kurland, Following Homology in Singularly Perturbed Systems, *J. Diff. Equ.*, 62(1986) 1-72.

[36] O.E. Lanford, A computer-assisted proof of the Feigenbaum conjectures, *Bull. AMS (N.S.)* **6**, 1982, 427-434.

[37] O.E. Lanford, Computer assisted proofs in analysis, *Physica A* **124**, 1984, 465-470.

[38] W.S. Massey *Homology and Cohomology Theory*, Marcel Dekker Inc., New York and Basel, 1978.

[39] Ch. McCord, The Connection Map for Attractor-Repeller Pairs, *Trans. AMS* 307(1988), 195-203.

[40] Ch. McCord, Mappings and Homological Properties in the Conley Index Theory, *Ergod. Th. and Dyn. Sys.* 8*(1988), 175-199.

[41] Ch. McCord, On the Hopf Index and the Conley Index, *Trans. AMS*, 313(1989), 853-860.

[42] K. Mischaikow, Existence of Generalized Homoclinic Orbits for One Parameter Families of Flows, *Proc. AMS* 103(1988), 59-69.

[43] K. Mischaikow, Transition Matrices, *Proc. Roy. Soc. Edinburgh* 112A(1989), 155-175.

[44] K. Mischaikow, Homoclinic Orbits in Hamiltonian Systems and Heteroclinic Orbits in Gradient and Gradient-Like Systems, *J. Diff. Equ.* 81(1989), 167-213.

[45] K. Mischaikow, Dynamic Phase Transitions: a Connection Matrix Approach, IMA Preprint Series no 584.

[46] K. Mischaikow, Travelling Waves for a Cooperative and a Competitive-Cooperative System, preprint CDSNS90-25.

[47] K. Mischaikow, A C-graph Approach for Studying the Dynamics of a System of Parabolic Equations, preprint.

[48] K. Mischaikow, On the Existence of Connecting Orbits for Scalar Delay Equations, preprint CDSNS90-35.

[49] K. Mischaikow and Y. Morita, Dynamics on the Global Attractor of a Gradient Flow Arising from the Ginzburg-Landau Equation, *JJIAM*, **11** (1994) 185-202.

[50] K. Mischaikow and M. Mrozek, Isolating neighborhoods and Chaos, *Jap. J. Ind. & Appl. Math.*, **12**, 1995, 205-236..

[51] K. Mischaikow and M. Mrozek, Chaos in Lorenz equations: a computer assisted proof, *Bull. Amer. Math. Soc. (N.S.)*, **33**(1995), 66-72.

[52] K. Mischaikow and M. Mrozek, Chaos in Lorenz equations: a computer assisted proof, Part II: details, preprint.

[53] K. Mischaikow, M. Mrozek and A. Szymczak, Chaos in Lorenz equations: a computer assisted proof, Part III: the classic case, in preparation.

[54] K. Mischaikow, G. Wolkowicz, Predator Prey with Group Defense, a Connection Matrix Approach, *Nonlin. Anal. Th. Meth. and Appl.*, 14(1990), 955-969.

[55] J.T. Montgomery, Cohomology of Isolated Invariant Sets under Perturbation , *J. Diff. Equ.* **13** (1973), 257-299.

[56] M. Mrozek, Index pairs and the Fixed Point Index for Semidynamical Systems with Discrete Time, *Fund. Mathematicae*, Vol. 133(1989), 177-192.

[57] M. Mrozek, The Cohomological Index of Conley Type for Multi-Valued Admissible Flows, *J. Diff. Equ.* 84.1(1990), 15-51.

[58] M. Mrozek, Leray Functor and the Cohomological Conley Index for Discrete Dynamical Systems, *Transactions of the American Mathematical Society* **318**, 1990, 149-178.

[59] M. Mrozek, Open Index Pairs and Rationality of Zeta Functions, *Ergod. Th. Dyn. Syst.* 10(1990), 555-564.

[60] M. Mrozek, The Morse Equation in Conley's Index Theory for Homeomorphisms, *Topology and its Appl.* 38(1991), 45-60.

[61] M. Mrozek, The Conley Index on Compact ANR's is of Finite Type, *Results in Math.* 18(1990), 306-313.

[62] M. Mrozek, Shape Index and Other Indices of Conley Type for Continuous Maps on Locally Compact Metric Spaces, *Fundamenta Mathematicae*, 145(1994),15-37.

[63] M. Mrozek, Topological invariants, multivalued maps and computer assisted proofs in dynamics, *Computers & Mathematics*, accepted.

[64] M. Mrozek, An algorithmic approach to the Conley index theory, in preparation.

[65] M. Mrozek, J. Reineck and R. Srzednicki, The Conley index over a base, in preparation

[66] M. Mrozek, J. Reineck and R. Srzednicki, The Conley index over a circle, in preparation.

[67] M. Mrozek, K.P. Rybakowski, Cohomological Conley Index for Continuous Maps on Metric Spaces, *J. Diff. Equ.* 90.1(1991), 143-171.

[68] M. Mrozek, K.P. Rybakowski, Bounded Solutions of Semilinear Parabolic Equations, *Proc. Roy. Soc. Edinburgh*, 117A(1991),305-315.

[69] M. Mrozek, K.P. Rybakowski, Conley Index of Difference Equations Approximating Differential Equations, *J. of Dynamics and Differential Equations* 4(1992),57-63.

[70] A. Neumaier, Th. Rage, Rigorous chaos verification in discrete dynamical systems, *Physica D* **67**, 1993, 327-346.

[71] Th. Rage, A. Neumaier, Ch. Schlier, Rigorous verification of chaos in a molecular model, *Physical Rev. E* **50**, 1994, 2682-2688.

[72] J. Reineck, The Connection Matrix and the Classification of Flows Arising from Ecological Models, Ph. D. Thesis, University of Wisconsin, Madison, 1985.

[73] J. Reineck, Travelling Wave Solutions to a Gradient System, *Trans. AMS* 307(1988), 535-544.

[74] J. Reineck, Connecting Orbits in One-Parameter Families of Flows, *Ergod. Th. Dyn. Sys.* 8*(1988), 359-374.

[75] J. Reineck, The Connection Matrix in Morse-Smale Flows I,II, *Trans. AMS*, to appear.

[76] J. Reineck, Continuation to Gradient Flows and the Conley Index, *Ergod. Th. Dyn. Sys.*, to appear.

[77] J.W. Robbin and D. Salamon, Dynamical systems, shape theory and the Conley index, *Erg. Th. and Dynam. Sys.* 8*(1988), 375-393.

[78] K.P. Rybakowski, On the Homotopy Index for Infinite Dimensional Semiflows, *Trans. AMS* 269(1982), 351-382.

[79] K.P. Rybakowski, The Morse Index, Repeller-Attractor Pairs and the Connection Index for Semiflows on Noncompact Spaces, *J. Diff. Equ.* 47(1983), 66-98.

[80] K.P. Rybakowski, *The Homotopy Index and Partial Differential Equations*, Springer-Verlag, Berlin Heidelberg 1987.

[81] K.P. Rybakowski, Nontrivial Solutions of Elliptic Boundary Value Problems with Resonance at Zero, *Annali Mat. Pura ed Appl.* (IV), CXXXIX(1985), 237-278.

[82] K.P. Rybakowski, An Index Product Formula for the Study of Elliptic Resonance Problems, *J. Diff. Equ.* 56(1985), 408-425.

[83] K.P. Rybakowski, On a Relation between the Brouwer Degree and the Conley Index for Gradient Flows, *Bull. Soc. Math. Belg. (B)*, 37(II)(1985), 87-96.

[84] K.P. Rybakowski, A Homotopy Index Continuation Method and Periodic Solutions of Second Order Gradient Systems, *J. Diff. Equ.* 65(1986), 203-212.

[85] K.P. Rybakowski, Some Recent Results in the Homotopy Index Theory in Infinite Dimensions, *Rend. Ist. Matem. di Trieste* XVIII(1986), 83-92.

[86] K.P. Rybakowski, On Critical Groups and the Homotopy Index in Morse Theory on Hilbert Manifolds, *Rend. Ist. Matem. Univ. di Trieste* XVIII(1986), 163-176.

[87] K.P. Rybakowski, E. Zehnder, A Morse Equation in Conley's Index Theory for Semiflows on Metric Spaces, *Ergod. Th. Dyn. Syst.* 5(1985), 123-143.

[88] D. Salamon, Connected Simple Systems and the Conley Index of Isolated Invariant Sets, *Trans. AMS* 291(1985), 1-41.

[89] D. Salamon, E. Zehnder, Flows on Vector Bundles and Hyperbolic Sets, *Trans. AMS* 306(1988), 623-649.

[90] E.H. Spanier, *Algebraic Topology*, McGraw-Hill Book Company, New York, 1966.

[91] A. Szymczak, The Conley index for discrete semidynamical systems, *Topol. Appl.*, submitted.

[92] A. Szymczak, The Conley index for decompositions of isolated invariant sets, *Fund. Math.*, to appear.

[93] W.C. Troy, The Existence of Steady State Solutions of the Kuramoto- Shivashin- sky Equation, *J. Diff. Equ.* 82(1989),269-313.

[94] J.R. Ward, Jr., Conley Index and Non-Autonomous Ordinary Differential Equations, *Results in Math.* 14(1988), 191-210.

[95] J.R. Ward, Jr., Averaging, Homotopy, and Bounded Solutions of Ordinary Differential Equations, *Diff. Int. Equ.* 3.6(1990), 1093-1100.

ON GEOMETRIC DETECTION OF PERIODIC SOLUTIONS AND CHAOS

R. Srzednicki

Jagiellonian University, Cracow, Poland

1 Introduction

In this note we consider a differential equation

$$\dot{x} = f(t, x) \tag{1}$$

where $f : \mathbb{R} \times M \to TM$ is a continuous time-dependent vector-field on a smooth manifold M. We assume that the Cauchy problem

$$x(t_0) = x_0 \tag{2}$$

associated to (1) has the uniqueness property. Fix some $T > 0$ and let the map $f(\cdot, x)$ be T-periodic for every $x \in M$. Examples of equations of that form arise, for example, in mechanical systems perturbed by periodic forces. One of the most important problems referred to (1) in that case is to determine the existence of its periodic solutions with the period being a multiplicity of T. Classical methods related to that problem are presented in [RM]. The purpose of this note is to describe a geometric approach to it, introduced in the paper [S2] (see also [S1] and [S4]). Actually, we present (without complete proofs) its minor modification and applications to detecting chaotic dynamics. Full exposition of the topics concerning chaos which are presented here is contained in the papers [S5] and [SW].

The basic idea of our approach is to find conditions for a set $W \subset \mathbb{R} \times M$ and f restricted to the boundary of W ensuring the existence of $(t_0, x_0) \in W$ such that the solution of the Cauchy problem (1), (2) is T-periodic. Those conditions should involve

---
*Supported by the KBN grant C/2375.

only information on geometry of W and $f|_{\partial W}$ – no à priori conditions on solutions of (1) outside of ∂W must be provided. That approach to detection of periodic solutions is partially motivated by a topological method of proving of the existence of a solution of the equation $G(x) = 0$ for a vector-field $G : \mathbb{R}^n \to \mathbb{R}^n$. Let $U \in \mathbb{R}^n$ be a closed bounded set, $U = \operatorname{cl} \operatorname{int} U$ with no zeros of G in ∂U. A sufficient condition for the existence of solution of $G(x) = 0$ in U is the nontriviality of the Brouwer degree $\deg(0, G, \operatorname{int} U)$. If U is an n-dimensional submanifold with boundary, by a result of Marston Morse (see [G]), the degree is completely determined by the Euler characteristic of U and the restriction of G to ∂U. If, moreover, G represents some suitable (but generic) behavior, the degree can be determined by the Generalized Poincaré Index Formula which involves only the Euler characteristics of U and some subsets of ∂U described by $G|_{\partial U}$, compare [P]. For smooth G, the existence of zeros of G in U is equivalent to the existence of stationary points of the flow ψ generated by the equation $\dot{x} = G(x)$ inside U (see [F] for a formula in the latter case in a purely topological setting). One can expect that the existence of periodic points inside U can also be determined by $G|_{\partial U}$ and some topological conditions imposed on U. However, except of 1-dimensional (where everything is trivial) and 2-dimensional phase space (where the Poincaré-Bendixson theory is valid), there is no general higher-dimensional result on that subject; for example there can be no periodic orbits inside a solid torus even if the vector-field restricted to its boundary is directed inward (this relates to the Seifert Conjecture, see [K] for a construction of a C^∞-counterexample). We make some additional assumptions an modifications; the first one which we propose is a rotation of the flow ψ. More precisely, we assume that the phase space of ψ has (up to a homeomorphism) the form $S^1 \times M$ (recall that M is a smooth manifold) and the velocity of ψ with respect to the S^1-variable is nonzero. We assume also that U is an isolating block (see [C] or [Sm]); in practice it means that there are no points of inward tangency of ψ with respect to the boundary of U. (The latter assumption is motivated, in some way, by the lack of periodic solutions for arbitrary small 2π-periodic resonant perturbations of the oscillator equation.) We make also some other assumptions on geometry of U, in particular (U, U^-), where U^- is the exit set, must be the total space of a pair of locally trivial bundles over S^1. At this moment we reformulate the problem: we identify S^1 with $\mathbb{R}/T\mathbb{Z}$ and, after a change of variables if necessary, we assume that the velocity of the rotation around S^1 is equal to 1. Thus the autonomous equation we consider is of the form

$$\begin{cases} \dot{t} = & 1 \\ \dot{x} = & f(t, x), \end{cases} \tag{3}$$

where f is described at the beginning of this text; the phase space of (3) can be either $S^1 \times M$ or its covering space $\mathbb{R} \times M$; our choice will depend on the current context. Moreover, there is a one-to-one correspondence between periodic orbits of ψ and periodic solutions of (1), hence instead of determining conditions for U and U^- in $S^1 \times M$ we will deal with the corresponding conditions for their lifts W and W^- in $[a, a + T] \times M$ for some $a \in \mathbb{R}$. Such a W will be called a *periodic isolating segment.*

its precise definition will be given in Section 2. In Section 3, an automorphism μ_W of homologies of (W, W^-) is defined. The main result, Theorem 1 states that if the Lefschetz number of μ_W is nonzero then the equation has a periodic solution whose graph over $[a, a + T]$ is contained in W. If the Lefschetz numbers determined by segments W and Z, with $Z \subset W$, are different then there is a periodic solution with graph over $[a, a + T]$ contained in W and not contained in Z (see Theorem 2).

Those theorems were successfully applied to problems on the existence of periodic solutions of non-autonomous planar polynomial equations in [S3], planar rational equations in [KS], and equations of higher order in [SS]. They can also be useful in detection of chaotic dynamics. Usual approach to that problem is based on proving the existence of suitable homoclinic or heteroclinic orbits (this leads to proofs of chaos for systems which are closely related to Hamiltonian ones, compare [W], or to computer-assisted proofs, like in [HHTZ]). Recently a new method, based on the discrete Conley index theory, was presented in [MM], however it still needs estimates on possible position of some solutions, which can be provided only by computer calculations. On the contrary to that methods, we are able to avoid calculations leading to global information on any concrete solution. As it can be seen below, suitable forms of periodic isolating segments guarantee the existence of a compact invariant set such that the Poincaré map restricted to that set is semiconjugated to the shift on two symbols and the counterimage (by the semiconjugacy) of any periodic point in the shift contains a periodic point of the Poincaré map. If that condition is satisfied, we call (1) *chaotic*. It follows in particular, that a chaotic equation has kT-periodic solutions for all $k = 1, 2, \ldots$. Obviously, one can modify that definition of chaos (and, consequently, results on its existence) by considering subshifts of finite type. Although chaos in that sense does not satisfy, in general, all the three axioms from the Devaney's definition (see [D]), it is stronger that the one defined in [MM], and guarantees positive topological entropy of the Poincaré map. The latter condition does not follow from the Devaney's axioms and its importance is indicated by some authors (compare Introduction in [GW]). We describe two different criteria for detecting chaos using periodic isolating segments. One of them, Theorem 3, comes from [S5] and the other, Theorem 4, from [SW] (its extensions will be given in [Wo]).

We give examples of planar differential equations to which the results presented in this note apply. More exactly, we indicate how to prove that the planar equation (written in complex numbers notation)

$$\dot{z} = e^{i\phi t}|z|^2 \bar{z} + 1$$

has a $2\pi/\phi$-periodic solution for $\phi \neq 0$,

$$\dot{z} = (1 + e^{i\phi t}|z|^2)\bar{z}$$

has a nonzero $2\pi/\phi$-periodic solution for $\phi \neq 0$ and is chaotic if $0 < \phi \leq 1/288$, and

$$\dot{z} = \frac{1}{2}e^{-i\phi t}z\left(\frac{1}{2}\phi(z+1) + e^{i\phi t}(\bar{z}+1)\right)\left(\frac{1}{2}\phi(z-1) + e^{i\phi t}(\bar{z}-1)\right)$$

is chaotic provided $\phi > 0$ is small enough.

2   Isolating segments

Let u be the *evolutionary operator* for the equation (1), i.e. $u_{(t_0,s)}(x_0) \in M$ is the value
of the solution of the Cauchy problem (1), (2) at the time $s \in \mathbb{R}$. u is continuous with
respect to all its variables, the domain of each map $u_{(t,s)}$ is open (possibly empty) in
M and two conditions are satisfied:

$$\forall t \in \mathbb{R} : u_{(t,t)} = \mathrm{id}_M,$$
$$\forall r, s, t \in \mathbb{R} : u_{(t,r)} = u_{(s,r)} \circ u_{(t,s)}.$$

If f is T-periodic with respect to t then

$$u_{(t+T,s+T)} = u_{(t,s)}$$

hence, in order to determine all T-periodic solutions of the equation (1), it suffices to
look for fixed points of the Poincaré map $u_{(s,s+T)}$ for any $s \in \mathbb{R}$.

Let $a, b \in \mathbb{R}$, $a < b$. Denote by $\pi_1 : [a,b] \times M \to [a,b]$ and $\pi_2 : [a,b] \times M \to M$ the
projections, and for a subset $Z \subset \mathbb{R} \times M$ and $t \in \mathbb{R}$ put

$$Z_t = \{x \in M : (t,x) \in Z\}.$$

Let (W, W^-) be a pair of subsets of $[a,b] \times M$. We call W an *isolating segment over*
$[a,b]$ (for the equation (1)) and W^- the *exit set of* W if:

(i) W and W^- are compact ENR's (i.e. Euclidean neighborhood retracts, compare
[Do]),

(ii) there exists a homeomorphism

$$h : [a,b] \times (W_a, W_a^-) \longrightarrow (W, W^-)$$

such that $\pi_1 = \pi_1 \circ h$,

(iii) for every $t \in [a,b)$ and $x \in \partial W_t$ there exists a $\delta \in (0, b-t]$ such that for every
$s \in (t, t+\delta)$ either $u_{(t,s)}(x) \notin W_s$ or $u_{(t,s)}(x) \in \mathrm{int}\, W_s$.

(iv) $W^- \cap ([a,b) \times M) = \{(t,x) \in W : t < b, \exists \delta > 0\ \forall s \in (t, t+\delta) : u_{(t,s)}(x) \notin W_s\}$.

Isolating segments over [a,b] can be constructed in the following way. Let $L^1, \ldots, L^s$:
$M \to \mathbb{R}$ be continuous functions (for some $s \geq 1$), let $g : \mathbb{R} \times M \to TM$ be a C^1 time-
dependent vector-field (in general, à priori not related to f) such that every solution
of the equation $\dot{x} = g(t,x)$ is defined in the whole interval $[a,b]$, and let v be the
evolutionary operator of that equation. Define $\Lambda^i : \mathbb{R} \times M \to \mathbb{R}$ by

$$\Lambda^i(t,x) = L^i(v_{(t,a)}(x)). \tag{4}$$

Proposition 1 *The set*

$$W = \{(t, x) \in [a, b] \times M : \forall i = 1, \ldots, s : \Lambda^i(t, x) \leq 0\},$$

is an isolating segment over $[a, b]$ *with the exit set*

$$W^- = \{(t, x) \in W : \exists i = 1, \ldots, r : \Lambda^i(t, x) = 0\}.$$

provided the following conditions are satisfied:

(j) *the set* $W_a = \{x \in M : \forall i = 1, \ldots, s : L^i(x) \leq 0\}$ *is compact,*

(jj) L^i *is* C^1 *in some neighborhood of the set* $\{x \in W_a : L^i(x) = 0\}$, *for each* $i = 1, \ldots, s$,

(jjj) *for every* $x \in W_a$ *the set* $\{\nabla L^i(x) : i$ *such that* $L^i(x) = 0\}$ *is linearly independent,*

(jw) *the inequalities*

$$\nabla \Lambda^i(t, x) \cdot (1, f(t, x)) > 0 \quad (\forall i = 1, \ldots, r, \ (t, x) \in W, \ \Lambda^i(t, x) = 0),$$

$$\nabla \Lambda^i(t, x) \cdot (1, f(t, x)) < 0 \quad (\forall i = r + 1, \ldots, s, \ (t, x) \in W, \ \Lambda^i(t, x) = 0)$$

hold.

Indeed, it is easy to see that (j) and (jjj) imply (i), (jw) implies (iii) and (iv), and h in (ii) can be defined as $h(t, x) = v_{(a,t)}(x)$.

An isolating segment W over $[a, a + T]$ is called *periodic* if

$$(W_a, W_a^-) = (W_{a+T}, W_{a+T}^-).$$

If f is T-periodic in t then the image of a periodic isolating segment over $[a, a + T]$ by the covering map $\mathbb{R} \times M \to S^1 \times M$ (where $S^1 = \mathbb{R}/T\mathbb{Z}$) is an isolating block for the flow generated by (3).

Example 1 Consider the planar equation

$$\dot{z} = \frac{1}{2}i\phi z + \beta e^{i\phi t}|z|^\alpha \bar{z} \tag{5}$$

where $z \in \mathbb{C}$, $\phi \neq 0$, $\alpha \geq 0$, and $\beta > 0$. It can be easily replaced by a system of two equations of real variables x and y if we put $z = x + iy$; actually, we do not distinguish between (x, y) and z. By Proposition 1, one can construct periodic isolating segments over $[0, 2\pi/\phi]$ for (5) as follows. Let $R > 0$ be a given number. For $(x, y) \in \mathbb{R}^2$ put

$$L^1(x, y) = \frac{1}{R^2}x^2 - 1 \quad L^2(x, y) = \frac{1}{R^2}y^2 - 1.$$

Let g be a planar vector-field given by $g(z) = \frac{1}{2}i\phi z$. The evolutionary operator v for the equation $\dot{z} = g(z)$ satisfies $v_{(t,0)}(z) = e^{-i\phi t/2}z$, hence (4) has the form

$$\Lambda_R^1(t, x, y) = \frac{1}{R^2}(x\cos(\frac{\phi}{2}t) + y\sin(\frac{\phi}{2}t))^2 - 1,$$

$$\Lambda_R^2(t, x, y) = \frac{1}{R^2}(x\sin(\frac{\phi}{2}t) - y\cos(\frac{\phi}{2}t))^2 - 1.$$

One can check that for $r = 1$ and $s = 2$ the condition (jw) is satisfied (for every R), hence

$$W(R) = \{(t, x, y) \in [0, 2\pi/\phi] \times \mathbb{R}^2 : \Lambda_R^i(t, x, y) \le 0, \; i = 1, 2\}$$

is an isolating segment for (5) and its exit set is given by

$$W(R)^- = \{(t, x, y) \in [0, 2\pi/\phi] \times \mathbb{R}^2 : \Lambda_R^1(t, x, y) = 0, \; \Lambda_R^2(t, x, y) \le 0\}.$$

If we suitably perturb the equation (5) then $W(R)$ still is an isolating segment for some values of R, as the following results show. Let $p : \mathbb{R} \times \mathbb{R}^2 \to \mathbb{R}^2$ be a smooth map.

Proposition 2 *If $\alpha > 0$ and*

$$\frac{p(t, z)}{|z|^{1+\alpha}} \longrightarrow 0 \text{ as } |z| \to \infty \text{ uniformly in } t \in [0, \frac{2\pi}{\phi}]$$

then there exists R_∞ such that for every $R \ge R_\infty$, $W(R)$ is an isolating segment with the exit set $W(R)^-$ for the equation

$$\dot{z} = \beta e^{i\phi t}|z|^\alpha \bar{z} + p(t, z). \tag{6}$$

Proposition 3 *If $\alpha \ge 0$ and*

$$\frac{p(t, z)}{|z|^{1+\alpha}} \longrightarrow 0 \text{ as } |z| \to 0 \text{ uniformly in } t \in [0, \frac{2\pi}{\phi}]$$

then there exists R_0 such that for every R satisfying $0 < R \le R_0$, $W(R)$ is an isolating segment with the exit set $W(R)^-$ for the equation

$$\dot{z} = \frac{1}{2}i\phi z + \beta e^{i\phi t}|z|^\alpha \bar{z} + p(t, z). \tag{7}$$

Remark 1 In the above propositions, $W(R)$ still remains an isolating segment with the exit set $W(R)^-$ for the given values of R if we perturb (6) (or (7), respectively) by adding a term $q(t, z)$ (for a smooth map q) with $|q(t, z)|$ small enough for each $(t, z) \in W(R)$. This is a consequence of the strong inequalities in (jw).

3   Periodic solutions in segments

Let W be an isolating segment over $[a, b]$. Define the homeomorphism

$$\tilde{h} : (W_a, W_a^-) \longrightarrow (W_b, W_b^-)$$

by $\tilde{h}(x) = \pi_2 h(b, \pi_2 h^{-1}(a, x))$ for $x \in W_a$, where h satisfies the condition (ii) in the definition of isolating segment. Geometrically, $\tilde{h}$ moves a point $x \in W_a$ to W_b along the arc $h([a, b] \times \{\pi_2 h^{-1}(a, x)\})$. A different choice of the homeomorphism h in (ii) leads to a map which is homotopic to $\tilde{h}$, hence the isomorphism

$$\mu_W = \tilde{h}_* : H(W_a, W_a^-) \longrightarrow H(W_b, W_b^-)$$

induced by $\tilde{h}$ in singular homology over the field of rational numbers $\mathbb{Q}$, is an invariant of the segment W.

In the reminder of this note we assume that f is T-periodic in t. If the segment W is periodic over $[a, a+T]$, then ν_W is an automorphism of $H(W_a, W_a^-)$ and its Lefschetz number is defined as

$$\text{Lef}(\mu_W) = \sum_{n=0}^{\infty} (-1)^n \text{tr } \tilde{h}_{*n}.$$

In particular, if $\mu_W = \text{id}_{H(W_a, W_a^-)}$ then $\text{Lef}(\mu_W)$ is equal to the Euler characteristic $\chi(W_a, W_a^-)$. The following theorems (up to slightly different notation) was proved in [S2] (see also [S1]):

Theorem 1 *If W is a periodic isolating segment over $[a, a + T]$ for the equation (1) and*

$$\text{Lef}(\mu_W) \neq 0$$

then (1) has a T-periodic solution $\phi : \mathbb{R} \to M$ such that $\phi(t) \in \text{int } W_t$ for every $t \in [a, a + T]$.

Theorem 2 *Let W and Z be two periodic isolating segments over $[a, a + T]$ for (1), $Z \subset W$. If*

$$\text{Lef}(\mu_W) \neq \text{Lef}(\mu_Z)$$

then there exists a T-periodic solution $\phi : \mathbb{R} \to M$ of (1) such that $\phi(t) \in \text{int } W_t$ for every $t \in [a, a + T]$ and $\phi(s) \notin Z_s$ for some $s \in [a, a + T]$.

In [S1] and [S2] the term periodic block (being the union of the copies of W translated by kT along the t-axis, $k \in \mathbb{Z}$) was used in the above statements. In the mentioned papers, and also in [KS], [S3], and [S4], there are many examples of equations satisfying Theorems 1 and 2. Below we shortly describe some of them.

Example 2 For every $\phi \neq 0$ the equation

$$\dot{z} = e^{i\phi t}|z|^2\bar{z} + 1 \tag{8}$$

has a $2\pi/\phi$-periodic solution.

Indeed, by Proposition 2, $W(R)$ is a periodic isolating segment for (8) if R is sufficiently large. Since $\text{Lef}(\mu_{W(R)}) = 1$, Theorem 1 implies the result.

Another proof of the existence of a periodic solution of (8) can be found in [Ma].

Example 3 For every $\phi \neq 0$ the equation

$$\dot{z} = (1 + e^{i\phi t}|z|^2)\bar{z} \tag{9}$$

has a nonzero $2\pi/\phi$-periodic solution.

Indeed, as in Example 2, $W(R)$ is a periodic segment for (9) if R is large enough. Put

$$Z(r) = [0, 2\pi/\phi] \times \{(x, y) \in \mathbb{R}^2 : |x| \leq r, |y| \leq r\},$$
$$Z(r)^- = [0, 2\pi/\phi] \times \{(x, y) \in \mathbb{R}^2 : |x| = r, |y| \leq r\}$$

$Z(r)$ is a periodic segment with the exit set $Z(r)^-$ for sufficiently small $r > 0$. Actually, we can put $R = 3$ and $r = \frac{1}{3}$. Obviously, we have $Z(\frac{1}{3}) \subset W(3)$. Since $\text{Lef}(\mu_{Z(r)}) = \chi(Z(r)_0, Z(r)_0^-) = -1$, the result follows by Theorem 2.

4   Chaotic dynamics

Let Σ_2 denote the set of bi-infinite sequences of two symbols and σ is the shift map (compare [W]). Recall that we already assumed that f is T-periodic in t. The equation (1) is called *chaotic* (or, more precisely, T-*chaotic*) provided there is a compact set $I \subset M$, invariant with respect to the Poincaré map $u_{(a,a+T)}$ (for some a), and a continuous surjective map $g : I \to \Sigma_2$ such that:

(k) $\sigma \circ g = g \circ u_{(a,a+T)}$, i.e. $u_{(a,a+T)}$ is semiconjugated to σ in the set I,

(kk) for every n-periodic sequence $s \in \Sigma_2$ its counterimage $g^{-1}(s)$ contains at least one n-periodic point of $u_{(a,a+T)}$.

It follows in particular that if (1) is T-chaotic then it has a periodic solution with basic period kT for every $k \in \mathbb{N}$ and the topological entropy of the Poincaré map is positive. Examples of chaotic equations can be given by classical results on detection of the existence of Smale's horseshoes for Poincaré maps. In particular, many equations T-periodic in t, considered in [W], are also kT-chaotic for some large number k.

We give two criteria for chaotic equations based on notions introduced in the previous sections. Let W and Z be periodic isolating segments over $[a, a + T]$ for (1). A proof of the following theorem will appear in [S5]:

Theorem 3 *Assume that:*

(A1) $(W_a, W_a^-) = (Z_a, Z_a^-)$,

(A2) $\exists s \in (a, a + T) : W_s \cap Z_s = \emptyset$,

(A3) $\exists n \in \mathbf{N} : H_n(W_a, W_a^-) = \mathbb{Q}, \ \forall k \neq n : H_k(W_a, W_a) = 0$.

Then (1) is T-chaotic.

The next theorem is the main result of [SW]:

Theorem 4 *Assume (A1) and:*

(B1) $Z \subset W$,

(B2) $\mu_Z = \mu_W \circ \mu_W = \mathrm{id}_{H(W_a, W_a^-)}$,

(B3) $\mathrm{Lef}(\mu_W) \neq \chi(W_a, W_a^-) \neq 0$.

Then (1) is T-chaotic.

Proofs of the above theorems are based on results of Section 3. One can construct segments of the form $V_0 \cup \tau_T(V_1) \cup \ldots \cup \tau_{kT}(V_k)$ over $[a, a + (k+1)T]$, where τ_s denotes the translation by $s \in \mathbb{R}$ along the t-axis and V_i is either equal to W or to Z, calculate possible values of the Lefschetz numbers of the corresponding homomorphisms, and apply Theorem 1 in the proof of Theorem 3 and, respectively, Theorem 2 in the proof of Theorem 4 in order to get the required periodic solutions.

5   Two chaotic equations

In this section we provide equations satisfying the criteria given in Section 4. As a consequence of Theorem 4 we get:

Corollary 1 *The equation (9) is $2\pi/\phi$-chaotic provided $0 < \phi \leq 1/288$.*

Segments satisfying hypothesis of Theorem 4 for (9) can be constructed as follows (see [SW] for details of that construction). Recall that the sets $W(R)$ and $Z(r)$ are defined in Example 1 and Example 3, respectively. Put $W = W(3)$. The segment Z can be obtained as a proper modification of $Z(\frac{1}{3})$ in order to get the concordance $(W_0, W_0^-) = (Z_0, Z_0^-)$. To this purpose we construct a (nonperiodic) segment P over $[0, 16]$ by $P_t = Z(3 - \frac{1}{6}t)$ and Q over $[2\pi/\phi - 16, 2\pi/\phi]$ by the mirror image of P with respect to the hyperplane $\{(t, x, y) : t = \pi/\phi\}$. Put $Z = P \cup Z(\frac{1}{3}) \cup Q$. It follows that $\mu_W = -\mathrm{id}$, $\mathrm{Lef}(\mu_W) = 1$, and $\chi(W_0, W_0^-) = -1$ hence Theorem 4 implies the result.

   The final result of this note can be proved using Theorem 3.

Corollary 2 *The equation*

$$\dot{z} = \frac{1}{2}e^{-i\phi t}z(\frac{1}{2}\phi(z+1) + e^{i\phi t}(\bar{z}+1))(\frac{1}{2}\phi(z-1) + e^{i\phi t}(\bar{z}-1)) \tag{10}$$

is $2\pi/\phi$-chaotic provided $0 < \phi \le \phi_0$, for some $\phi_0 > 0$.

We explain briefly how to prove that result – details will be given in [S5]. By $g(t, z)$ denote the right-hand side of (10). In order to find proper segments for (10), observe first that

$$g(t, z) = \frac{1}{2}e^{i\phi t}|z|^2\bar{z} + \{\text{terms of order} \le |z|^2\} + \phi \cdot \{\text{terms of order } |z|^3\},$$

hence, by the form of considered terms, Proposition 2 and Remark 1, $W(R)$ is a segment for (10), with the exit set $W(R)^-$, if R is large and ϕ is small enough. Similarly, since

$$g(t, z) = (\frac{1}{2}i\phi(z-1) + e^{i\phi t}\overline{z-1}) + \{\text{terms of order} \ge |z-1|^2\} + $$
$$\phi \cdot \{\text{terms of order } |z-1|\},$$

Proposition 3 and Remark 1 imply that

$$U(r) = \{(t, x, y) : (t, x-1, y) \in W(r)\}$$

is also a segment for (10) with the exit set

$$U(r)^- = \{(t, x, y) : (t, x-1, y) \in W(r)^-\}$$

provided r and ϕ are sufficiently small. Fix such R and r, assume additionally that $r < 1$ and $R > 1+r$. Choose a (large) number $\omega > 0$ and a (small) number $\epsilon > 0$ such that the sets

$$P = \{(t, x, y) : t \in [-\omega, 0], \ |x| \le R, \ |y| \le \frac{r-R}{\omega}t + r\},$$

$$Q = \{(t, x, y) : t \in [0, \epsilon], \ \frac{1-r+R}{\epsilon}t - R \le x \le \frac{r+1-R}{\epsilon}t + R, \ |y| \le r\}$$

are isolating segments for the equation

$$\dot{z} = \frac{1}{2}z(\bar{z}+1)(\bar{z}-1),$$

which satisfy hypotheses of Proposition 1 (especially, the condition (jw) is of our interest), such that their exist sets are given by

$$P^- = \{(t, x, y) : t \in [-\omega, 0], \ |x| = R, \ |y| \le \frac{r-R}{\omega}t + r\},$$

$$Q^- =$$
$$\{(t,x,y) : t \in [0,\epsilon], x = \frac{1-r+R}{\epsilon}t - R \text{ or } x = \frac{r+1-R}{\epsilon}t + R, |y| \le r\}.$$

Define W, the first of the segments considered in Theorem 3, as follows:

$$W_t =$$
$$\begin{cases} U(r)_t \text{ for } t \in [\epsilon, \frac{\pi}{\phi} - \epsilon], \\ W(R)_t \text{ for } t \in [\frac{\pi}{\phi} + \omega, \frac{2\pi}{\phi} - \omega], \\ \{(t,x,y) : (t, (x - \frac{t}{\epsilon})\cos(\frac{\phi}{2}t) + y\sin(\frac{\phi}{2}t) + \frac{t}{\epsilon}, (x - \frac{t}{\epsilon})\sin(\frac{\phi}{2}t) - \\ \qquad\qquad\qquad y\cos(\frac{\phi}{2}t)) \in Q\} \text{ for } t \in [0,\epsilon], \\ \{(t,x,y) : (t - \frac{2\pi}{\phi}, x\cos(\frac{\phi}{2}t) + y\sin(\frac{\phi}{2}t), x\sin(\frac{\phi}{2}t) - y\cos(\frac{\phi}{2}t)) \in P\} \\ \qquad\qquad\qquad \text{for } t \in [\frac{2\pi}{\phi} - \omega, \frac{2\pi}{\phi}], \end{cases}$$

and for $t \in [\pi/\phi - \epsilon, \pi/\phi + \omega]$, the set W_t is defined in an analogous way as in $[0, \epsilon]$ and $[2\pi/\phi - \omega, 2\pi/\phi]$. It can be proved (after precise formulation of the above construction) that if ϕ is small enough (actually, much smaller than $1/\omega$), then W is a periodic segment for (10) over $[0, 2\pi/\phi]$; it sourrounds points with the x-coordinate equal to 1. By the symmetric argument, we can construct a periodic segment Z, surrounding points with the x-coordinate -1, which coincides with W over the interval $[\pi/\phi, 2\pi/\phi]$. Moreover, W_ϵ is disjoint from Z_ϵ and homologies of (W_0, W_0^-) are equal to the homologies of S^1, hence Theorem 3 implies the result.

References

[C] C. C. Conley, Isolated Invariant Sets and the Morse Index, CBMS vol. 38, Amer. Math. Soc., Providence, RI 1978.

[D] R. Devaney, Chaotic Dynamical Systems, Addison-Wesley, New York 1989.

[Do] A. Dold, Lectures on Algebraic Topology, Springer-Verlag, Berlin, Heidelberg, New York 1972.

[F] R. E. Feßler, A generalized Poincaré index formula, J. Differential Equations 115 (1995), 304-323.

[G] D. H. Gottlieb, A de Moivré like formula for fixed point theory, Contemp. Math. 72 (1988), 99-105.

[GW] E. Glasner, B. Weiss, Sensitive dependence on initial conditions, Nonlinearity 6 (1993), 1067-1075.

[HHTZ] B. Hassard, S. P. Hastings, W. C. Troy, J. Zhang, A computer proof that the Lorentz equations have "chaotic" solutions, Appl. Math. Lett. 7 (1994), 79-83.

[KS] T. Kaczynski, R. Srzednicki, Periodic solutions of certain planar rational equations with periodic coefficients, Differential Integral Equations 7 (1994), 37-47.

[K] K. Kuperberg, A smooth counterexample to the Seifert conjecture, Ann. Math. 140, 3 (1994), 723-732.

[Ma] J. Mawhin, Periodic solutions of some complex-valued differential equations with periodic coefficients, in: G. Lumer et al (editors), Partial Differential Equations. Contributions to the conference held in Han-sur-Lesse, Math. Res. 82, Akademie-Verlag, Berlin 1993, 226-234.

[MM] K. Mischaikow, M. Mrozek, Chaos in the Lorenz equations: a computer-assisted proof, Bull. Amer. Math. Soc. 32, 1 (1995), 66-72.

[P] C. C. Pugh, A generalized Poincaré index formula, Topology 7 (1968), 217-226.

[RM] N. Rouche, J. Mawhin, Ordinary Differential Equations. Stability and Periodic Solutions, Pitman, Boston, London, Melbourne 1980.

[SS] S. Sędziwy, R. Srzednicki, On periodic solutions of certain n-th order differential equations, J. Math. Anal. Appl. 196 (1995), 666-675.

[Sm] J. Smoller, Shock Waves and Reaction-Diffusion Equations, Springer-Verlag, New York, Heidelberg, Berlin 1983.

[S1] R. Srzednicki, A geometric method for the periodic problem in ordinary differential equations, Séminaire d'Analyse Moderne, No. 22, Université de Sherbrooke 1992, 1-110.

[S2] R. Srzednicki, Periodic and bounded solutions in blocks for time-periodic non-autonomous ordinary differential equations, J. Nonlinear Anal. - Theory, Meth. Appl. 22, 6 (1994), 707-737.

[S3] R. Srzednicki, On periodic solutions of planar polynomial differential equations with periodic coefficients, J. Differential Equations 114 (1994), 77-100.

[S4] R. Srzednicki, A geometric method for the periodic problem, in: V. Lakshmikantham (editor), Proceedings of the First World Congress of Nonlinear Analysts, Tampa, Florida, August 19-26, 1992, Walter de Gruyter, Berlin, New York 1996, 549-560.

[S5] R. Srzednicki, On chaotic dynamics inside isolating blocks, in preparation.

[SW] R. Srzednicki, K. Wójcik, A geometric method for detecting chaotic dynamics, preprint, Jagiellonian University, Kraków 1996.

[W] S. Wiggins, Global Bifurcation and Chaos. Analytical methods, Springer-Verlag, New York, Heidelberg, Berlin 1988.

[Wo] K. Wójcik, On detecting periodic solutions and chaos in ODEs, in preparation.

The participants to the CISM School on
"Non Linear Analysis and Boundary Value Problems for Ordinary Differential Equations"

Printed in the United States
By Bookmasters